Pesticides Identification
at the Residue Level

A symposium sponsored by the Division of Pesticide Chemistry of the American Chemical Society at the Joint Conference of the Chemical Society of Canada and the American Chemical Society at Toronto, Ontario, Canada, May 26–27, 1970.

Francis J. Biros

Symposium Chairman

ADVANCES IN CHEMISTRY SERIES **104**

AMERICAN CHEMICAL SOCIETY

WASHINGTON, D. C. 1971

Coden: ADCSHA

Library of Congress Catalog Card 70–164408

ISBN 8412–0119–6

PRINTED IN THE UNITED STATES OF AMERICA

Advances in Chemistry Series

FOREWORD

ADVANCES IN CHEMISTRY SERIES was founded in 1949 by the American Chemical Society as an outlet for symposia and collections of data in special areas of topical interest that could not be accommodated in the Society's journals. It provides a medium for symposia that would otherwise be fragmented, their papers distributed among several journals or not published at all. Papers are refereed critically according to ACS editorial standards and receive the careful attention and processing characteristic of ACS publications. Papers published in ADVANCES IN CHEMISTRY SERIES are original contributions not published elsewhere in whole or major part and include reports of research as well as reviews since symposia may embrace both types of presentation.

CONTENTS

PREFACE

The newly formed Division of Pesticide Chemistry of the American Chemical Society has as one of its central objectives the task of increasing and diffusing knowledge relating to the chemical, physical, and biological properties of the chemicals used to control pests of plants, animals, and man. In this regard, the Division promotes research in a number of important areas of pesticide chemistry ranging from the synthesis of newer, more effective pest control materials and techniques to scientific appraisals of the fate and persistence of these compounds in the environment. Directly pertinent to virtually every area of study is the development and utilization of methods for organic and inorganic pesticide analysis.

The tremendous achievements which have been accomplished by the judicious use of pesticides in increased world-wide food production and the control of vector-borne diseases are very evident. In addition, pesticidal materials contribute to the effective control of undesirable species of pests including insects, plants, bacteria, fungi, etc., and assist in the nutrition, growth, and reproduction of certain desirable species. However, because of widespread use in these applications, pesticides have proliferated intensely in the biosphere and thus have significantly contributed to problems of environmental pollution. For example, the effects of these chemicals on public health and the survival of species of fish and wildlife are two specific areas of critical concern. The attention which has been focused in recent years on the occurrence of residues of pesticides and their metabolites, as well as other industrially significant chemicals in the environment, is demonstrated by the numerous reports of various official and unofficial committees of inquiry which have considered this problem and have reported the results of large numbers of assays of pesticide residues in such diverse substrates as human and animal tissues, food, plants, water, soil, and air.

A majority of these residue determinations are made at the parts per million and parts per billion levels which require on the one hand the availability of analytical techniques of remarkable sensitivity, accuracy, and integrity in recording proper and reliable data and on the other, skilled interpretation, experience and knowledge of the capabilities and limitations of these techniques, and the application of suitable confirmatory methods of identification. In addition to these problems of detection and identification of infinitely small amounts of these materials, the complexity of the analysis of pesticide residues is further compounded by the prolific development of efficient new classes of pesticides which undergo environmental decomposition to a series of breakdown products by a number of mechanisms including metabolism, photodecomposition, hydrolysis, and/or oxidation. Each of these transformation products must be monitored to an extent equal to that of the parent material, particularly in those cases where the toxicological characteristics have not been ascertained.

The continued utilization of many "classical" pesticidal materials of demonstrated ubiquitousness and persistence and the presence of unknown or unsuspected industrial chemical pollutants other than pesticides in the environment are other areas of concern confronting public health officials and residue chemists alike. Thus, the analytical chemist is faced with a continuing responsibility to develop, evaluate, and improve the confidence of the techniques which are used to identify and quantitatively monitor these chemicals in human, animal, and environmental samples.

Within the framework of this philosophy, this symposium is concerned with several topic areas pertinent to the problems associated with the identification and confirmation of pesticide residues at submicrogram levels. The purpose of this symposium is to survey the present status of the classical physical, chemical, and biochemical techniques employed for residue analysis; to evaluate critically the more recently applied, relatively unexploited techniques which have become available for the detection, quantitation, and characterization of pesticide residues; and finally, to examine and discuss the current needs and problems, stimulate and encourage greater research activity, and point the way to future developments in the analysis of pesticide residues.

The four general topic areas to be discussed in the papers of the symposium include: philosophical aspects of ultramicroanalysis, instrumental techniques, microchemical methods, and biological assay methods. The aspects considered in most detail obviously will involve those encompassing instrumental techniques. Individual emphasis will be placed on gas-liquid chromatographic detectors, infrared and ultraviolet spectrophotometry, paper and thin-layer chromatography, mass spectrometry, and neutron activation analysis. Other papers will discuss recently developed techniques employing chemical and photochemical conversion of pesticides into derivatives useful for the confirmation of the parent residue at picogram and nanogram concentration levels. The utility of biological assay methods for the identification of residues will also be explored particularly from the point of view of enzymatic and immunological techniques. These methods have been shown to possess high sensitivity and their usefulness in certain applications is well established. Finally, the philosophical aspects of ultramicroanalysis will be discussed with emphasis on a survey of the specific problems associated with identification at submicrogram levels.

F. J. Biros

Perrine Primate Research Branch
Division of Pesticide Chemistry and Toxicology
Food and Drug Administration
Department of Health, Education and Welfare
Perrine, Florida 33157
August 1970

1

Possible Limits of Ultramicro Analysis

GUNNAR WIDMARK

Institute of Analytical Chemistry, University of Stockholm,
Stockholm 50, Sweden

This paper discusses the possible limits of sensitivity to be reached by modern methods of residue analysis. Development trends in analytical chemistry seem to follow two main paths, both marked by increasing sensitivity. The first path leads toward the traditional goal of quantitation, characterized by the strong demand for simplicity. The second path is hopefully leading to identification, with the need for complex information. Combination instruments employing gas chromatography (GC) and mass spectrometry (MS) can produce information for identification at high sensitivity, especially when using on-line data acquisition systems (DAS). A further increase in sensitivity and selectivity is foreseen when the combination of instruments is extended to cover the GC/MS/DAS/computer. The ultimate limits of ultramicro analysis are illustrated by a "map of tracer cosmos."

During the past two decades there has been a remarkable increase in sensitivity of chemical analysis, especially in pesticide residue analysis. It is from this field more than any other that high sensitivity figures such as ppm, ppb, and ppt have been brought to the public by the news media. Unfortunately, the public has had few opportunities to understand the dimensions of these figures; illustrating the biological significance of analytically high sensitivity figures is obviously an even more difficult task.

Ultimate Sensitivity at Ultramicro Analysis

It is reasonable to ask the analyst if there exists an ultimate possible limit of detection of small quantities in chemical analysis. As in most cases associated with pesticide chemistry, there are only complex answers to this very simple question.

What is actually questioned is the limit of sensitivity at which useful analytical results can be obtained. Thus, any answer would demand a definition of the concept of usefulness.

The experienced analyst knows that the extreme ultimate of sensitivity of one molecule (or one atom after decomposition) per sample unit is not foreseen for practical analytical work. Nevertheless, the early school of nuclear physicists were able to detect single elementary particles when using very simple instrumentation. The tremendous improvements in instrumentation since then would certainly make any type of single molecule detectable, but not in a mixture and never in an unknown sample given to the analyst.

In the sample of the old physicist, the emitted particle differed sufficiently from the background to be detected—when not absorbed! Although less sensitively detected, the same holds true to some extent for electron-capturing compounds, such as DDT, when present in an ordinary biological sample. The main difference is that there are only a few elementary particles to be considered, whereas there is a vast number of electron-capturing compounds. Thus, the ECD response will not be interpretable in proper chemical terms.

The main factor hampering an increase of sensitivity in chemical analysis—such as residue analysis—is the large number of compounds possibly present in an ordinary-size sample—*e.g.*, 1 gram. The large number of compounds makes it likely that many of the compounds will have chemical properties too close for differentiation by ordinary methods of detection and separation. However, because most modern schemes of analysis contain one or several separation and concentration steps, there is no more reason to make a nomenclature differentiation between ultramicro analysis and high sensitivity analysis. In both cases, high sensitivity detection is asked for, and by concentration steps the compounds will finally be detectable in a very small sample.

Tracer Cosmos

Although an enormous number of compounds are constituents of any 1-gram sample, as mentioned, this number is never unlimited. Approaching the lowest levels of concentration, where only a few orders of magnitude of each molecule are present, the number of possible constituents is such that the use of the name "tracer cosmos" is justified, emphasizing the analytical difficulties foreseen to reach the lower levels.

In Figure 1, an attempt is made to illustrate by the heavy lines, the sides of the triangles, the maximum number of compounds possibly

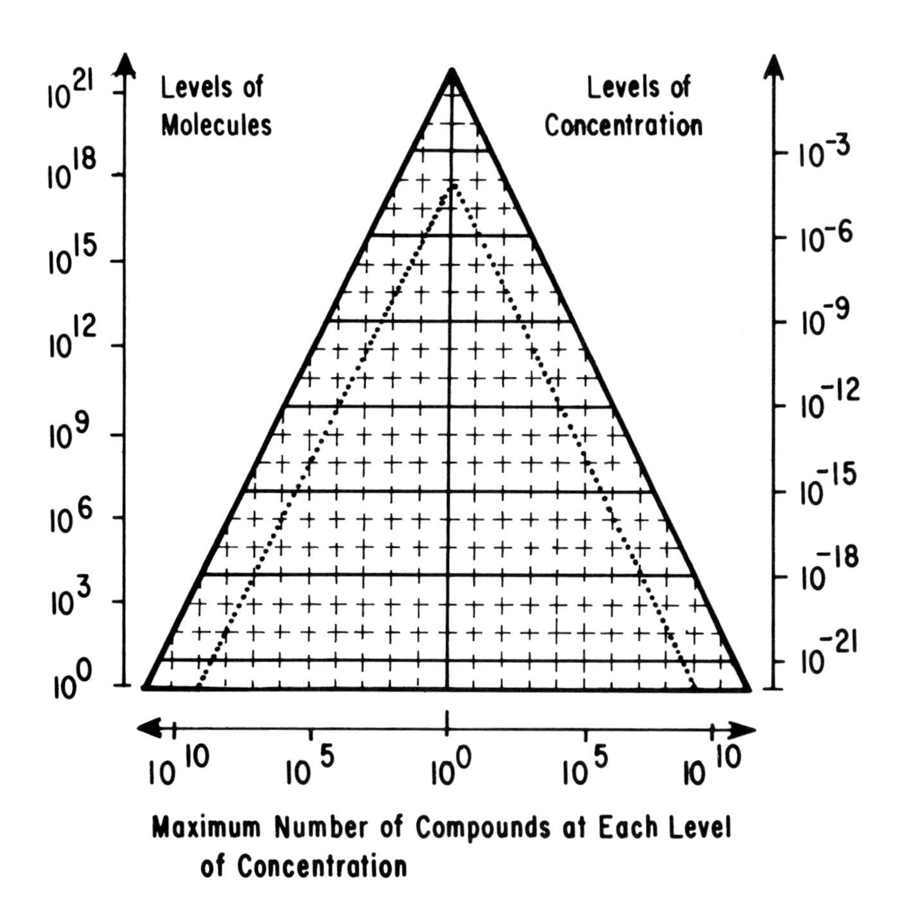

Figure 1. Map of tracer cosmos

present on each decimal level of concentration. In this figure, it is assumed that all compounds have a molecular weight of 60. This figure is too low, but 600 is believed to be far too high. Units of concentration now in use at pollution studies are given on the axes.

The dotted lines of Figure 1 refer to the assumption that we are dealing with a 1-gram sample of 99% purity, and that the amount of impurity is equally divided on each of the possible decimal levels of concentration. Naturally, to meet this assumption, there are not enough materials available for the 1% or for the 0.1% level. However, one is unlikely to find from natural sources an even distribution of contaminants like this, or one at which there are the same number of compounds on each level. The same holds true for the other extreme, that all the trace constituents are present on the same level of concentration. It seems more realistic to assume—excluding a detailed discussion—that impurities are grouped as "islands" on the triangular map of the tracer cosmos; presumably, the islands will be densest at the top. Depending on the chemical nature of these islands—lipid or hydrophilic—they will move upwards or downwards on their respective maps when brought in contact with a system of reversed character. Thus, a water sample will lose part of its dissolved lipid materials to suspended particles having a fatty surface.

Particles able to absorb lipid materials are always present in natural waters and aid in the transport of the lipid materials—including some technical compounds, *e.g.*, DDT—to living organisms, thus causing accumulation. Moreover, it is assumed that the fatty surface of the particles protects partly lipid materials from chemical degradation. However, it would be of great interest to know more about the reactions occurring in water at very low concentration levels where the highly reactive products of H_2O–O_2 equilibrium are to be found—*e.g.*, the hydroxy radical. Knowledge from this field would probably indicate that few organic compounds are likely to survive at the lowest levels of concentration.

The short discussion in the preceding paragraphs illustrates the usefulness of pesticide residue analysis at much lower levels of concentration than those presently available. As indicated by Table I, there are a number of other fields where analytical methods of improved sensitivity will be needed.

Table I. Fields Requiring More Sensitive Analysis

Studies of disappearance of technical chemical products in nature; *e.g.*, plastic, paint, pesticides, drugs (total fate studies; accumulation, *e.g.*, at the end of food chains)

Identification of unknowns found in nature (condition for legislation)

Studies of naturally-occurring compounds at ppm–ppb levels

Studies of main components of individual cells

New ways to determine degree of toxicity (at normally subtoxic levels)

Quality control and monitoring

Early warning systems (*e.g.*, health control)

Limits of Detection; Quantitation and Identification

Sensitivity limits must be improved for both quantitation and identification of pesticide residues. Although the former method is more sensitive than the latter, until recently most progress has been made in improving the sensitivity of quantitative analysis. Since there is a dependence between the two methods, sensitive methods of identification have

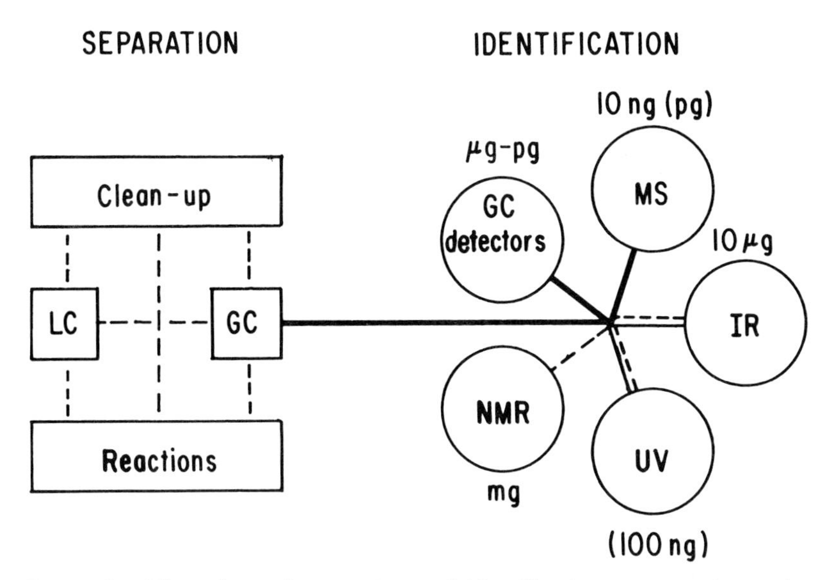

Figure 2. Flow chart of separation and identification processes in pesticide residue analysis. Solid line indicates that on-line combinations are available; thick line indicates that they are in common use; dotted line, manipulations.

become even more desirable. Figure 2 demonstrates the connection between the methods of separation and of identification; differences in sensitivities are indicated.

Quantitation. The most easily interpreted part of a quantitative residue analysis is the recording of a detector response found linear to concentration when checked by standard solutions. It has to be verified also that the response of sample and standard solution with a specific method is obtained at the same analytical position—*e.g.*, on a gas chromatogram. The use of blanks will demonstrate the degree of influence of interfering materials and thus exhibit the detection limit of the recording system. However, because of the fundamental limitation of any monodetector system, only considerable knowledge of the sample will make a quantitative residue analysis reliable.

A chemical analysis is not only the recording of a detector signal but a very complex matter. In Table II, the extent of total residue analysis is sketched, and at each step errors can be made which will influence the analytical result. When serious, these errors will convert the recorded analytical result into nonsense. When less important, these errors will only lower the over-all sensitivity of the analysis. Thus, the sensitivity limit obtainable by the detector alone is in reach only when the entire scheme of the analysis is successfully performed.

Table II. The Analytical Process

Planning before the actual analysis in collaboration with, *e.g.*, biologists

Discussions with biologists of the purpose of analysis
Discussion of the type of sample to be chosen
Pilot tests; use of simplified models
Suggestions—criticism
Small series are analyzed
Agreements on purpose, size, economy, etc.

Performance of analysis

Field sampling
Packing—storing
Transportation
Storing at the lab
Mechanical treatment
Extraction, etc.
Cleaning up
Chemical reactions
Preparation of final solution, standard, and blanks

Instrumental Analysis

Recording of response
Checks

Report

Calculations
Written report
Statistical treatment
Bookkeeping

After report

Discussions with biologists
Criticism
Corrections
Improvement of analytical method

A further negative influence on the possible limit of detection of residue analysis is given by the type of residue under investigation. There is a marked difference in detector sensitivity in analysis of, *e.g.*, carbamates and DDT-type pesticides. The general difficulties of each class have to be considered, as is demonstrated in Table III.

Identification. The fundamental limitation of gas chromatography in identification studies is that by this method, one can only state non-presence of a known compound which is available as a standard in a considerably pure form. Thus, gas chromatography is an excellent

Table III. Objects of Analytical Investigation

Compound(s) sought

Compounds slightly changed by:

Isomerization	Autoxidation
Polymerization	Hydrolysis
Photoreaction	

Metabolites (might vary with ecosystem)

Reactions in the ecosystem
e.g., Methylation

Induced changes in the ecosystem
e.g. Hormone effects, "Hit and run" effects

Possible Complications

Isomers and homologs

Related compounds
a) used for the same purpose
b) not used for the same purpose—*e.g.*, PCB

Contamination; impurities

method to demonstrate that a known pesticide is not present at a given level of concentration—*e.g.*, the one accepted by the authorities. However, no positive information as to the identity of the peak-forming compound is obtainable by ordinary gas chromatography. Some information might be collected using a two-detector system, but only if one of the detectors gives a response interpretable in chemical terms. As critical examination will show, the gas chromatographic detectors now used as compliments to ECD are imperfect in this respect. In general, more information is gained by using alternative methods of separation, as shown at the left of Figure 2. These operations will also serve a confirmatory purpose.

The great advantage of mass spectrometry over other methods of identification is that the response is given by integer mass units, and thus the response will be more apt for chemical interpretation than is valid for other types of detectors. Since the mass spectrometer, when combined with a gas chromatograph, will serve as a multidetector, the computerized data acquisition systems now being introduced on the market will improve our ability to identify most compounds separated in a reasonably pure form by gas chromatography. An ECD chromatogram should not be used alone as an indication of a successful separation. In some samples, such as sewage sludge, there might be a considerable overlap by compounds of low-capturing ability. Thus, FID checks should be made before starting identification studies on a mass spectrometer.

A severe obstacle in present identification studies of assumed pesticide residues is that the mass spectrometer is normally a less sensitive detector than ECD. When the mass spectrometer is set on a single mass number, there is a gain in sensitivity but also an obvious loss in selectivity. Recording of only a small—but significant—part of the mass spectrum while the gas chromatographic fraction emerges is an improvement for some identification studies, but column bleeding might give unexpected difficulties. However, it is probable that a desired gain in sensitivity will be achieved soon by instrumental improvements of the mass spectrometer, mainly in ionization. In Table IV, the losses which are assumed at the various sites of commercially available spectrometers are tabulated. This table also indicates that the mass spectrometer is obviously superior in sensitivity to other instrumental techniques. More-

Table IV. Detection Levels of Combined Instruments, GC/MS

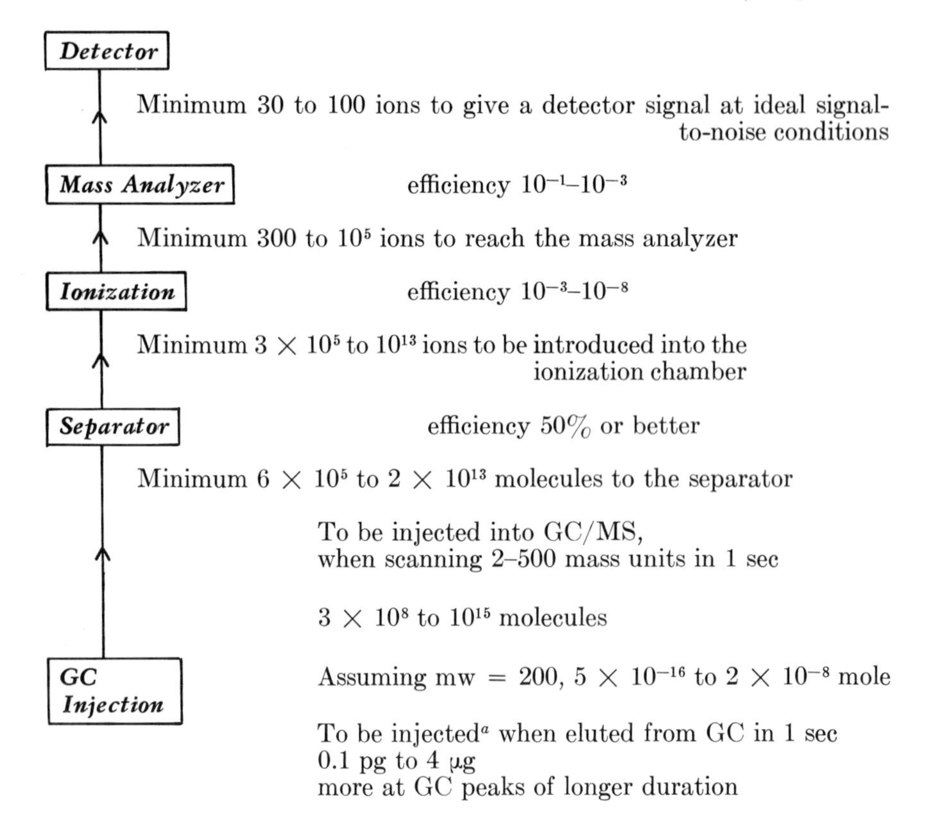

[a] In practice, it is necessary to prepare sample solutions 10–100 times more concentrated.

over, gas chromatographic peaks of short duration give an increase in sensitivity. This is one of the factors that favor the use of capillary columns for future residue studies.

Most analysts now using mass spectrometry alone or in combination with gas chromatography will not be able to utilize all the information obtainable at a series of mass spectrometric scans. This is mainly because of imperfect recording devices which are unable to accept all details in the rapid flow of signals. Some analysts complain of being completely drowned in paper spectra. We have been lucky enough to have one of the early data acquisition systems capable of producing instantaneously compensated and normalized mass spectra in digital form (*1, 2*). This system (on-line) has gradually been extended by a recent connection to a small computer; thus, we can conveniently use any system of recorded data as long as this operation can be programmed. The rapid changes between different analytical programs will be facilitated by an external disc memory now being installed (*3*). Our new system will then operate as demonstrated by the flow sheet in Figure 3. However, despite the use of all the electronics, the success of an analysis will still depend mainly

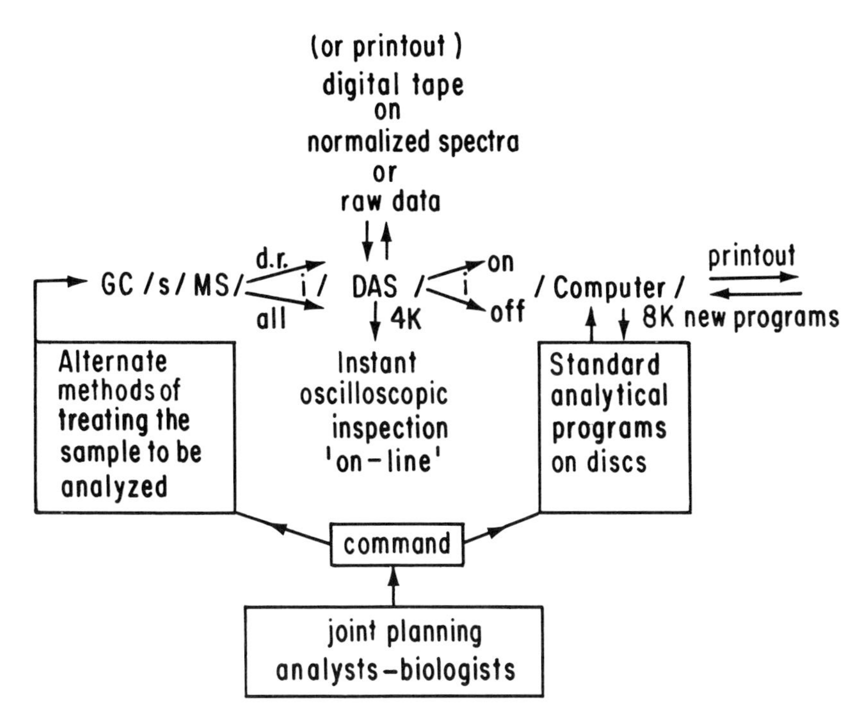

Figure 3. Flow sheet of a computerized combination gas chromatograph (GC) and mass spectrometer (MS); DAS = data acquisition system, s = separator, i = interface, d.r. = data reduction

on the condition of the gas chromatographic column and the proper handling of the sample.

The programming of a comprehensive data system integrated to a system of analytical instrumentation will certainly be a very exciting task. However, this will demand specific information in a form which is useful for programming from all the scientists involved in the study or responsible for parts of the study, as indicated by Tables II and III. Present difficulties in achieving this type of collaboration might remain as a limiting factor for improved limits of residue analysis detection and other similar types of ultramicro analysis.

Literature Cited

(1) Bergstedt, L., Widmark, G., *Chromatographia* (1969) **2,** 529.
(2) *Ibid.*, (1970) **3,** 59.
(3) Bennet, P., Bergstedt, L., Widmark, G., to be published.

RECEIVED August 24, 1970.

2

Chemical Derivatization Techniques for Confirmation of Organochlorine Residue Identity

W. P. COCHRANE and A. S. Y. CHAU

Analytical Services Section, Plant Products Division, Department of Agriculture, Ottawa, Ontario

Misidentifications in organochlorine pesticide residue analysis by thin-layer and gas chromatography occur as a result of interferences from co-extracted pesticide compounds and naturally-occurring products as well as external contamination. A useful technique for confirmation of residue identity is chemical derivatization. Addition, oxidation, rearrangement, dechlorination, reduction, and dehydrochlorination are the most commonly used procedures. A table of confirmatory tests for organochlorine residues is given for the proper choice and application of a derivatization reaction. The sensitivity of these methods ranges from 0.01 to 0.1 ppm in terms of the parent pesticide present in a 10-gram field-treated sample. The mode of formation of some chlordan derivatives is discussed with reference to the mechanism of reaction and structure of the parent compound.

In recent years much concern has risen, especially in official regulatory circles, about the problem of misidentification or uncertain identification in pesticide residue analysis. Since the introduction of the electron-capture detector (EC) in 1960 (*1*) and its rapid exploitation for the determination of organochlorine residues by gas-liquid chromatography (GLC) (*2*) the combined EC–GLC system has become, from 1963, the most commonly used end-method for quantitative pesticide residue analysis. It was quickly discovered that even after the application of the more common clean-up techniques (*3, 5, 4*) EC–GLC interferences occurred not only from peak-overlap of the various pesticides themselves (*6, 7*) but also from extraneous contamination—*e.g.*, the laboratory or technique itself (*8*)—and naturally-occurring components co-extracted from the sample.

For example, artifacts (*9*) having similar GLC responses to *o,p'*- and *p,p'*-DDE on both DC-200 and QF-1 stationary phases were obtained by using polyethylene wash bottles (*10*). Similarly, both elemental sulfur (*11*) and di-*n*-butylphthalate (*12*) have been identified as interfering with aldrin identification. More recently, naturally-occurring components of green plant material have been reported with similar retention times, again on DC-200 and QF-1 columns, to dieldrin (*13*). This list of interfering responses can be extended to include previously unknown derivatives or metabolites such as those found with heptachlor and heptachlor epoxide. From technical chlordan-treated cabbage the dehydrochlorinated product of *trans*-chlordan, namely 2-chlorochlordene (Figure 1), was found to have EC–GLC and thin-layer chromatographic (TLC) characteristics similar to heptachlor (*14*). To date, no suitable GLC column has been obtained that will successfully separate these two compounds, and further it is not known whether the 2-chlorochlordene originated as a minor constituent of technical chlordan or is, in fact, a *trans*-chlordan metabolite. A heptachlor epoxide artifact (*15*), observed during *cis*- and *trans*-chlordan feeding experiments on rats (*16*), has been subsequently identified as the closely-related compound 1,2-dichlorochlordene epoxide (*17*), which possesses one more chlorine atom in the 2-endo-position than heptachlor epoxide itself (Figure 1). This new chlordan metabolite has also been observed in milk (*18*). Of more general interest is the group of industrial compounds known as the polychlorinated biphenyls (PCB) which, on the commonly used GLC columns, interfere with practically all organochlorine insecticides (*19, 20, 21*). With the advent of numerous shortened methods of residue analysis in which the final result is obtained *via* extraction procedures incorporating GLC without prior clean-up, the situation may become even worse.

It has already been stated that "no one method can identify an unknown residue with absolute certainty" (*22*) and in this context four parameters have been suggested "from which the confirmation of identity of a residue can be inferred with reasonable assurance" (*23*). Two of these parameters are the EC–GLC retention time (on a given stationary phase) before and after chemical reactions. The pesticides or metabolites concerned are converted before injection to derivatives with different retention times from the parent compounds and also from other common pesticides that may be present. The other two parameters are R_f values (TLC or PC) or *p*-values (*24*) and insecticidal activity.

Although TLC has been widely accepted as a source of additional information in corroboration of residue identity (*9*), the use of chemical

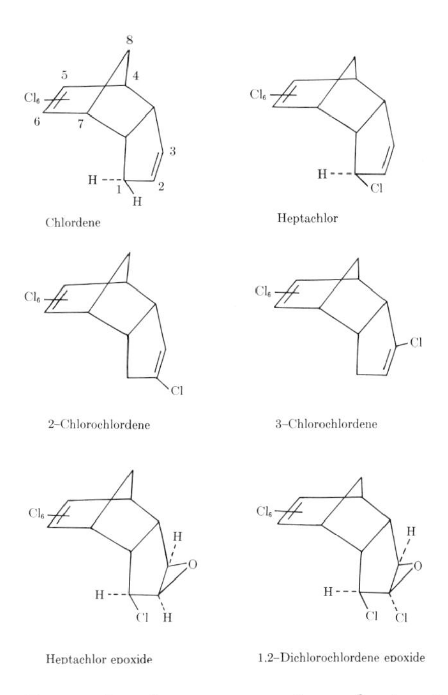

Figure 1. Structures and numbering for some chlordan compounds and isomers

derivatization has been a relatively recent innovation owing primarily to a change in application. Prior to 1965, chemical methods were used as an integral part of the extraction/clean-up procedures to remove or modify the co-extracted material (*25*). Since that time, the number of publications dealing with residue confirmation has gradually increased and in particular the organochlorine residues have been converted to GLC-responsive derivatives by many methods including addition, oxidation, rearrangement, dechlorination, and dehydrochlorination reactions. In many instances, the GLC chromatogram obtained after reaction has been "cleaned-up" of interfering or background components that were present in the original sample. Confirmatory tests for several groups of organochlorine pesticides have now been established in terms of specific or general reagents and the mechanistic pathways by which they are derivatized.

The DDT Group

Saponification has been by far the most commonly used method for both the identification and, in many instances, quantitation of DDT and related compounds. Derivative techniques have been applied to biologi-

Table I. Confirmatory Tests for Pesticide

Pesticide	*Reaction Utilized*
DDT	a) Dehydrochlorination b) Dechlorination (of *p,p'*–isomer)
DDE	Oxidation
DDD	Dehydrochlorination
Methoxychlor	Dehydrochlorination
Aldrin	a) Addition — Cl_2 / Br_2 / *tert*–BuOCl b) Epoxidation
Dieldrin	Epoxide — cleavage / rearrangement / acetylation
Endrin	Epoxide rearrangement Dechlorination
Endosulfan	Sulfite reduction

[a] In ppm of parent pesticide based on a 10-gram sample.
[b] 6% QF–1 + 4% DC–11 on Chromosorb W. Column operating at 200°C.

cal extracts containing *p,p'*-DDT, *p,p'*-DDD, dicofol, Perthane, *p,p'*-methoxychlor, and in many instances their respective *o,p'*-isomers. While the majority of these DDT-related compounds are converted to their respective olefins (Figure 2), dicofol yields *p,p'*-dichlorobenzophenone. These confirmatory reactions have been evaluated on standard insecticide solutions (*25, 26, 27*), vegetable (*28, 29*) and animal tissue (*27, 30*), and soil extracts (*31*), and the corresponding dehydrochlorinated derivatives display R_t (retention time) values substantially different from the parent compounds. In some cases, misleading results can be obtained by the less than astute analyst because of peak-overlap of the expected derivative with an organochlorine pesticide residue already present but unaffected by the alkali conditions used. Using a column containing Chromosorb W coated with 4% SE-30 + 6% QF-1, the derivatives DDMU (Figure 2) and *o,p'*-DDE, from *p,p'*-DDD and *o,p'*-DDT, respectively, overlap with heptachlor epoxide which is unaffected under mild basic conditions (*28*). Similarly, *p,p'*-DDE and dieldrin coincide using DC-200 as stationary phase (*31*).

Although potassium *tert*-butoxide (*t*-BuOK), sodium methylate ($NaOCH_3$), sodium ethylate ($NaOC_2H_5$), alcoholic potassium hydroxide, and sodium hydroxide solutions have been used, the utilization of the

Residues; The DDT Group and Some Cyclodienes

Limit of Detectability [a]	*Pesticide Interference* [b]	*Reference*
0.02	DDE	*27*
0.03	*p,p'*–DDD	*32*
0.05	many	*29*
0.03	–	*27*
0.1	*p,p'*–DDT	*27*
0.03	*p,p'*–DDD	*30*
0.04	–	*41*
0.04	–	*14*
0.01	Dieldrin	*42*
0.01	Endrin	*30*
0.05	Endrin	*36*
0.04	–	*39*
0.04	Dieldrin	*36*
0.04	–	*40*
0.02	*o,p'*–DDE	*53*

stronger nucleophilic reagents—*e.g.*, *t*-BuOK—can simultaneously identify DDT and analogues and some cyclodiene insecticides of the chlordan series (*14*).

Normally, *p,p'*-DDT occurs together with its metabolite *p,p'*-DDE, which can interfere with dehydrochlorination confirmation unless it has been previously removed, for example, by TLC (Table I). One chemical

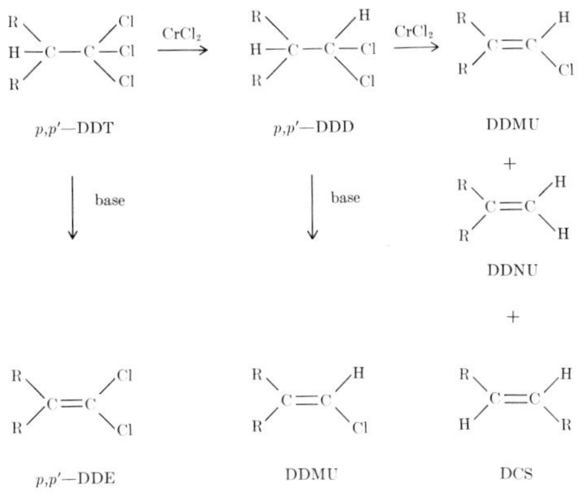

Figure 2. Reaction products of p,p'-*DDT and* p,p'-*DDD with basic and aqueous chromous chloride reagents (where R = Cl*⟨◯⟩*–)*

reaction that circumvents this additional step is dechlorination with chromous chloride ($CrCl_2$) solution. Chromous chloride reacts preferentially with the *p,p'*-isomer to give primarily *p,p'*-DDD (Figure 2) after only 45 minutes at 60°C (Table I). Prolonged reaction (Figure 2) further converts *p,p'*-DDD to DDNU, DDMU, and *trans-p,p'*-dichlorostilbene (DCS) (*32*). On a 4% DC-11 + 6% QF-1 on Chromosorb W column, *p,p'*-DDE and DCS have identical retention times. Also, the further photoisomerization product of DCS, the corresponding *cis*-isomer (*33*), DDNU, and heptachlor have the same R_t value (Figure 3). Reductive dechlorination with $CrCl_2$ does not appear to be common to all DDT-related compounds. For example, *o,p'*-DDT reacts fairly quickly and *p,p'*-methoxychlor not at all. Unpredictably, preferential dechlorination pathways have also been observed in the degradation of DDT and analogues in some biological systems but not in others (*34*).

Chromic acid oxidation has been employed for the quantitation of *p,p'*-DDE (*35*) and can be successfully applied to its confirmation on the residue scale (Table I). Since this reaction is applicable to other DDT-related compounds and metabolites, the possible sources of interference hazards in a DDT-containing biological extract are many.

When incorporated into regulatory residue procedures, these methods work satisfactorily with various plant and animal extracts down to levels of 0.02–0.1 ppm, depending on the pesticide concerned (Table I).

The Aldrin Group

Taking advantage of the observation that endrin is readily rearranged to a pentacyclic ketone by mineral acids (*36*) and also boron trifluoride (BF_3) (*37*), a number of confirmatory tests have been devised. Concentrated sulfuric acid (H_2SO_4) achieves complete conversion in 10–15 minutes at room temperature. Similarly, hydrobromic acid (HBr), chromic acid (CrO_3), and hydrochloric acid (HCl) under various conditions have been utilized for the positive identification of endrin found in soil, vegetable, forage crop, and fat extracts. Although the use of mineral acids is convenient and quick, dieldrin also reacts under similar conditions. With BF_3 in methanol, concentrated H_2SO_4, or perchloric acid, dieldrin is rearranged to a "dieldrin ketone," the exact structure of which is still not known. Using concentrated HCl in ethanol, the corresponding aldrin chlorohydrin has been identified as the derivative produced by epoxide cleavage (*38*). Presumably HBr yields the corresponding bromohydrin product and, if glacial acetic acid is present, the corresponding bromoacetate. In one instance (*21*), concentrated H_2SO_4 has been used for the detection of small amounts of the more volatile compounds—*e.g.*, aldrin and lindane—by "removal" of the oxygen-containing residues such

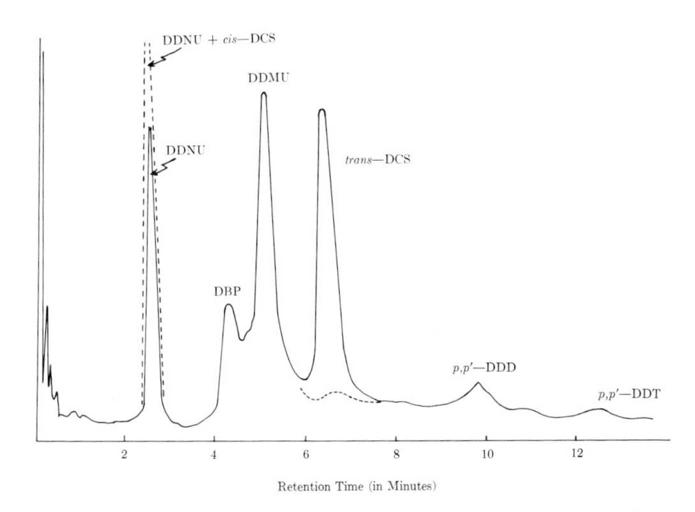

Figure 3. Chromatogram showing the products obtained from reaction of p,p'*-DDT with $CrCl_2$ after 24 hours at 60°C (———) and subsequent standing in fluorescent light or sunlight (- - - -)*

as dieldrin, endrin, and heptachlor epoxide. Unlike endrin and dieldrin, the disappearance of heptachlor epoxide is not readily explained since this residue is appreciably resistant to acid treatment.

Irrespective of the acidic reagent employed, if a DC-200 or mixed DC-11/QF-1 column is used in the final GC analysis, incomplete resolution of the endrin and dieldrin derivative peaks is observed (*36, 38*). This overlap problem has been overcome by using either 15% QF-1 and 10% DC-200 as stationary phase for on-column resolution or an acetylation procedure to give easily GC resolvable derivatives when endrin and dieldrin residues occur together. In the former case, this particular column was used to separate endrin ketone and the aldrin chlorohydrin; in the latter case an acetic anhydride–sulfuric acid reagent (*36, 39*) produced the expected endrin ketone and a dieldrin diacetate derivative having significantly different R_t values.

An alternative confirmatory test for endrin in the presence of dieldrin has been achieved by preferential reductive dechlorination. Although both endrin and dieldrin react with $CrCl_2$, endrin is substantially more labile (Figure 4) to this particular reagent to give a pentachloroketone which has a different retention time from the more common organochlorine pesticides (*40*).

Selective halogenation of aldrin to the corresponding dichloride or dibromide constitutes one approach to its confirmation (*30, 41*). However, careful removal of residual traces of chlorine or bromine must be performed to give GC-interpretable derivative-containing extracts. Also,

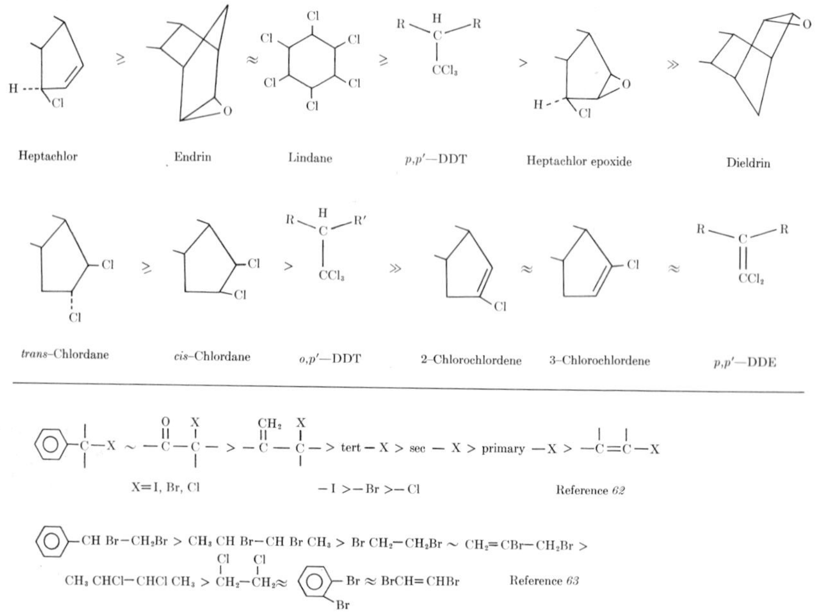

Heptachlor ≥ Endrin ≈ Lindane ≥ p,p'—DDT > Heptachlor epoxide ≫ Dieldrin

trans-Chlordane ≥ *cis*-Chlordane > o,p'—DDT ≫ 2-Chlorochlordene ≈ 3-Chlorochlordene ≈ p,p'—DDE

X=I, Br, Cl — tert — X > sec — X > primary —X

—I >—Br >—Cl Reference *62*

C_6H_5—CH Br—CH_2Br > CH_3 CH Br—CH Br CH_3 > Br CH_2—CH_2Br ∼ CH_2=CBr—CH_2Br > CH_3 CHCl—CHCl CH_3 > CH_2Cl—CH_2Cl ≈ $C_6H_4Br_2$ ≈ BrCH=CHBr Reference *63*

Figure 4. Relative order of reactivity of organochlorine pesticides and some halogenated compounds to Cr^{2+} reagents

dichloroaldrin may interfere with *p,p'*-DDD which is unaffected by the chlorination conditions. Addition to the unhindered double bond of aldrin has also been accomplished with *tert*-butyl hypochlorite (*tert*-BuOCl) under various conditions. Using *tert*-butyl alcohol as solvent, only aldrin undergoes reaction to give the corresponding chloro-*tert*-butyl ether,

Table II. The

Pesticide	*Reaction Utilized*
Heptachlor	a) Allylic — acetylation, hydroxylation, dechlorination b) Addition c) Epoxidation
Heptachlor epoxide	Epoxide rearrangement
cis- and *trans*-Chlordan	Dehydrochlorination
Nonachlor	a) Dechlorination b) Dehydrochlorination

whereas in glacial acetic acid both heptachlor and aldrin give their respective chloroacetate derivatives (*14*). The most convenient method, however, is the expoxidation of aldrin to dieldrin by peracids or CrO_3. Although peracetic (*42*), performic, perbenzoic, and monoperphthalic acids (*14*) are applicable for this epoxidation, *m*-chloroperbenzoic acid will probably be the reagent of choice because of its availability, even though it has been pointed out that background EC–GLC interferences may be an added complication (*43*).

The sensitivity range of these confirmatory chemical tests for aldrin, dieldrin, and endrin is 0.01 to 0.05 ppm in terms of the parent pesticides in field-treated samples (Table I).

The Chlordan Group

A variety of methods has been devised for the confirmation of heptachlor residues (Table II). The presence in the heptachlor molecule (Figure 1) of a reactive allylic chlorine atom has been the basis of three confirmatory tests based on its ease of replacement. Reaction with a silver acetate–glacial acetic acid mixture produced 1-acetoxychlordene which, with the GLC conditions used, had a retention time close to heptachlor epoxide (*44*). Of the common organochlorine pesticides, only heptachlor reacted quantitatively. Endrin reacts to a small extent with the glacial acetic acid to give a secondary endrin ketone peak. When the reaction of heptachlor with silver salts was extended to silver carbonate in aqueous alcohol, 1-hydroxychlordene was obtained which can easily be converted to the more volatile and GC-responsive silyl ether. Unfortunately, this silyl ether has a R_t identical to aldrin. With silver carbonate, hepta-

Chlordan Group

Limit of Detectability	*Pesticide Interference*	*Reference*
0.01	–	*44*
0.01	Aldrin	*44*
0.01	Chlordene	*45*
0.03	–	*14*
0.05	Heptachlor epoxide	*29*
0.01	–	*52*
0.01	–	*14*
0.01	*cis*-Chlordan	*53*
0.01	Aldrin	

chlor was the only chlorinated pesticide to undergo modification, and to date this reaction is the most specific confirmatory test observed for any of the compounds investigated. Since both the silver salt reactions involve refluxing for a minimum of 30 minutes, a more convenient laboratory technique is the use of $CrCl_2$ solution for the allylic dechlorination of heptachlor to chlordene (*45*). This is easily accomplished by reaction at 50°–60°C for 30 minutes. Together with the major reductive dechlorination pathway, two minor products, a dimer and 1-hydroxychlordene, are obtained probably by bimolecular reduction and hydroxyl insertion respectively, when the reaction is carried out in aqueous media (Figure 5). Under such conditions, it can also be postulated that 1-hydroxychlordene is formed by hydrolysis of heptachlor as has been previously observed (*46, 47*). On the residue scale, these secondary products present no observable GC interference. Chlordene has been obtained by dechlorination of heptachlor by bacteria and subsequent microbial epoxidation to chlordene epoxide in soils (*48*); therefore, interference can occur in such samples depending upon the extent of chlordene degradation.

Two further tests for heptachlor involve addition to the sterically hindered double bond. Normally, chlorine does not add to heptachlor unless an initiator, such as antimony pentachloride, is present. Heptachlor is resistant to epoxidation by peracids but both addition and epoxidation can readily be achieved chemically by *tert*-BuOCl/HOAc and CrO_3 oxidation, respectively (Table II). For the addition of *tert*-BuOCl to pro-

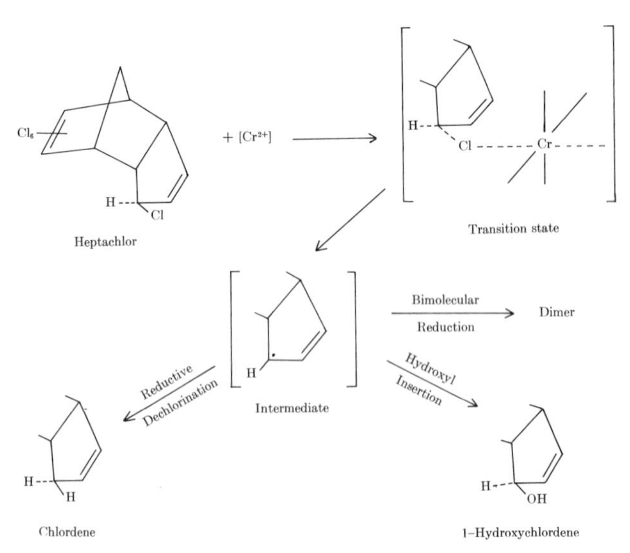

Figure 5. Reaction pathways of heptachlor with aqueous $CrCl_2$ solution

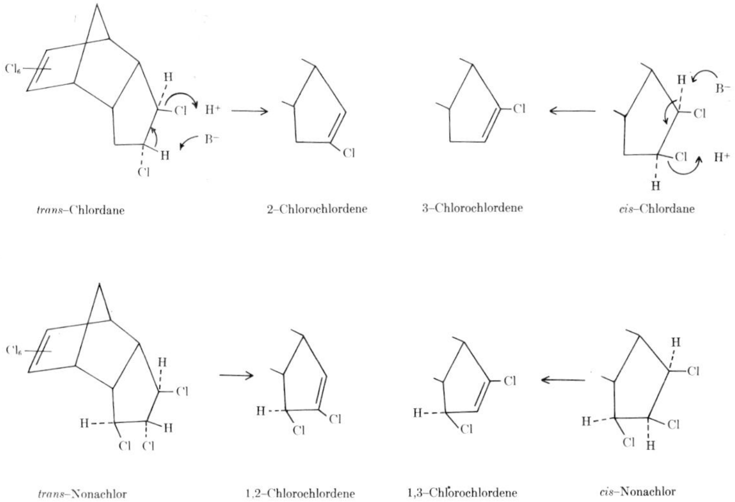

Figure 6. Dehydrochlorination mechanisms and products of isomeric chlordans and nonachlors

duce a single derivative peak which has been identified as the corresponding chloroacetate, an excess of glacial acetic acid is required (*14*).

Confirmatory oxidation of heptachlor, using chromic acid, produces its toxic metabolite heptachlor epoxide (*29, 49, 50*). Attempted identification of heptachlor in the presence of its epoxide produced no appreciable increase in the peak height of heptachlor epoxide because of its further substantial degradation to acidic products. Therefore this test should only be employed in the absence of heptachlor epoxide. This can either be achieved during clean-up using a Florosil column for their separation or subsequently using TLC. In combination, heptachlor and its epoxide can be simultaneously confirmed with strong basic reagents. Under these conditions both *cis*- and *trans*-chlordan as well as *trans*-nonachlor, which occur in varying amounts in technical chlordan, can also be confirmed. Using *tert*-BuOK/*tert*-BuOH, heptachlor produces 1-hydroxychlordene whereas heptachlor epoxide undergoes rearrangement to the secondary alcohol, 1-hydroxy-3-chlorochlordene (*51*). Both derivatives can be converted to more GC-responsive compounds either by silylation or acetylation of the allylic hydroxyl groups. With basic reagents, *cis*- and *trans*-chlordan are dehydrochlorinated to 3- and 2-chlorochlordene, respectively (*14, 52*) (Figure 6). A parallel observation was made when the dehydrochlorinated products of *cis*- and *trans*-nonachlor were shown to be 1,3- and 1,2-dichlorochlordene, respectively (*53*). With the two trans-isomers, dehydrochlorination occurs by the more difficult cis-elimination owing to the expected loss of the 2-exo-protons. On

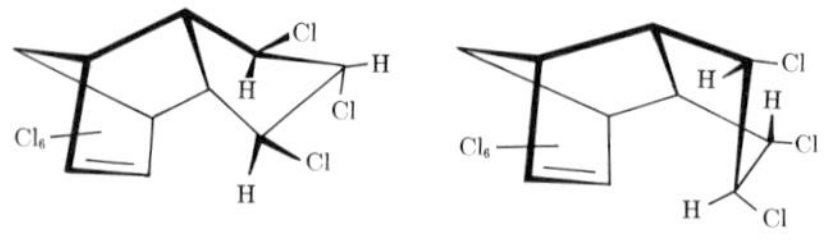

trans-Nonachlor *cis*-Nonachlor

Figure 7. Stereochemical structures of two isomeric nonachlors

the other hand, the two cis-isomers lose HCl rapidly by the normal trans-elimination mechanism but unexpectedly by the removal of the 3-endo-protons rather than the "apparently less hindered" 2-endo-protons. A ready explanation of these facts is possible if the stereochemical configurations of the cis- and trans-isomers are taken into account. In the trans-configuration, the 2-chlorine atom is in the endo-position and as such is influenced by the π-electron cloud of the ClC:CCl group. The configuration that will be assumed is one in which the 2-endo chlorine atom is in an environment as far removed from the π-electron cloud as possible. Conversely, when the 2-chlorine atom is in the exo-position, the 2-endo-proton is attracted to the π-electrons and so the cis-isomer assumes a different conformation. The stereochemical structures for *cis*- and *trans*-nonachlor illustrate this point (Figure 7). In this connection, it has been reported that *trans*-chlordan does not undergo photoisomerization while *cis*-chlordan gives a single 2-endo-cross-linked product in 70% yield (*54*).

At the residue level, all the chlordan compounds listed in Table II can be simultaneously confirmed using a *tert*-BuOK/*tert*-BuOH reagent. Silylation or acetylation of the hydroxy products do not affect any accom-

Table III. Some

Parent Pesticide	*Metabolite*
Heptachlor	a) Chlordene
	b) 1-Hydroxychlordene
	c) 1-Hydroxy-2,3-epoxy-chlordene
trans-Chlordan	2-Chlorochlordene
cis-Chlordan	3-Chlorochlordene
cis- and *trans*-Chlordan	1,2-Dichlorochlordene epoxide
Endrin	Photo-endrin
Endosulfan	Endosulfan diol

panying dehydrochlorinated products. The level of detectability for a 10-gram sample extract is 0.01 ppm. Lower levels can be reached by use of a larger sample together with appropriate clean-up techniques.

Miscellaneous

The α- and β-isomers of endosulfan can be reduced with lithium aluminium hydride in tetrahydrofuran to furnish the same endosulfan diol (*55*) which has been observed as one of its metabolic products (*56*). Depending on whether acetylation or silylation is used to make the diol more GC-responsive, lower limits of detectability are of the order 0.03 or 0.02 ppm, respectively (Table I).

For lindane residues, reliance must still be placed on the parameters afforded by R_t, R_f, *p*-values, and appearance in the appropriate fraction on column clean-up since no positive chemical confirmatory test is known. "Negative" indications to its identity can be obtained by its disappearance under basic or reducing conditions (*e.g.*, $CrCl_2$) while it remains intact in acidic or oxidizing media.

Negative type chemical reactions also have been employed for the identification of the polychlorinated biphenyls (PCB) when they occur together with organochlorinated pesticide residues. Nitration, using a 1:1 mixture of concentrated H_2SO_4 and concentrated nitric acid, of a PCB/pesticide mixture results in the disappearance of the more common organochlorine residues, especially the DDT group, while PCB and lindane are unaffected (*21*). It has also been observed that toxaphene is unaffected by such nitration reagents (*57*) while with alkaline treatment (*58*) this mixed isomer pesticide is converted to a characteristic pattern useful for identification purposes.

Organochlorine Metabolites

Group Utilized	*Derivative*
i Allylic hydrogen	1-Bromochlordene
ii Double bond	Chlordene epoxide
i Allylic hydroxy	Silyl ether
ii Double bond	Chloroacetate; epoxide
i Hydroxyl	Silyl ether
ii Epoxide	Trihydroxy chlordan
Double bond or *gem*-Dichloro group	Epoxide or hexachloro
Chloro epoxide or *gem*-Dichloro group	Chloro acetate or heptachloro
gem-Dichloro	Pentachloro
Hydroxyl	Acetate or silyl ether

During investigations on the development of confirmatory tests for the chlordans, endrin, and endosulfan, some metabolites or possible metabolites were also included. Again many of the procedures already described for acetylation, epoxidation, addition, dechlorination, etc., can be applied easily to the confirmation of identity of these degradation products (Table III). For both 3-chlorochlordene and the photoisomerization product of endrin (*i.e.*, endrin ketone), the *gem*-dichloro groups, present in the hexachlorocyclopentene moiety, are monodechlorinated with $CrCl_2$ to yield the respective pentachloro derivatives. This method of confirmation is generally applicable to a large number of organochlorine pesticides. The rate and extent of reaction depends upon both the availability of labile chlorine and the structure of the pesticide concerned. The general order of reactivity is as outlined in Figure 4. For comparison, two other series of halogen-containing compounds are included.

Conclusion

The above brief review illustrates that chemical derivatization techniques have been used extensively for the confirmation of identity of organochlorine residues. In most instances, the lower limits of detectability of the derivatives are substantially lower than the established tolerance values for the parent compounds. Taken in conjunction with the many other modes of derivatization—*e.g.*, during or after gas chromatography (*59*)—the analyst has a vast array of modification procedures at hand to aid in residue identification. They can be employed for residues in soil, biological, fat, and nonfat extracts and can be successfully extended, especially the more specific tests, to the identification of cross-contaminants in pesticide formulations and also fertilizer mixtures. So far, these latter two cases have only been a fringe area of application (*60, 61*).

It is a desirable practice to identify the structure of the derivative used in confirmatory identification as this will help the analyst not only comprehend the mechanisms and structures of the pesticides involved but also the limitations of a particular technique. Also, the more embracing the analyst or researcher's knowledge of the extent and probable sources of error and inherent short-comings of an analytical test method, the more chance of finding a successful solution and thereby avoiding the possibility of misidentification. Present trends appear to indicate that the continuing need for methods of identification will ensure that chemical derivatization tests will become applied routinely for the confirmation of residues of not only the organochlorines but also organophosphates, carbamates, and other classes of pesticides.

Literature Cited

(1) Lovelock, J. E., Lipsky, S. R., *J. Am. Chem. Soc.* (1960) **82,** 431–3.
(2) Goodwin, E. S., Goulden, R., Reynolds, J. G., *Analyst* (1961) **86,** 697–709.
(3) McKinley, W. P., McCully, K. A., *Can. Food Ind.* (1967) **38** (2), 47–53.
(4) *Ibid.,* (1967) **38** (3), 56–60.
(5) Schechter, M. S., Getz, M. E., *J. Assoc. Offic. Anal. Chemists* (1967) **50,** 1056–61.
(6) Burke, J., Holswade, W., *J. Assoc. Offic. Anal. Chemists* (1964) **47,** 845–59.
(7) *Ibid.,* (1966) **49,** 374–85.
(8) Lamar, W. L., Goerlitz, D. F., Law, L. M., ADVAN. CHEM. SER. (1966) **60,** 187–99.
(9) McCully, K. A., *World Rev. Pest Control* (1969) **8,** 59–74.
(10) Mestres, R., Berthes, F., Priu, F., *Trav. Soc. Pharm. Montpellier* (1966) **26,** 93–8.
(11) Pearson, J. R., Aldrich, F. D., Stone, A. W., *J. Agr. Food Chem.* (1967) **15,** 938–9.
(12) Sissons, D. J., Telling, G. M., Usher, C. D., *J. Chromatog.* (1968) **33,** 435–49.
(13) Glotfelty, D. E., Caro, J. H., *Anal. Chem.* (1970) **42,** 282–4.
(14) Chau, A. S. Y., Cochrane, W. P., *J. Assoc. Offic. Anal. Chemists* (1969) **52,** 1092–100.
(15) Polen, P. B., Hester, M., Benziger, J., *Bull. Environ. Contam. Toxicol.,* in press.
(16) Polen, P. B., *J. Assoc. Offic. Anal. Chemists* (1969) **52,** 300.
(17) Schwemmer, B., Cochrane, W. P., Polen, P. B., *Science* (1970) **169,** 1087.
(18) Lawrence, J. H., Barron, R. P., Chen, J-Y. T., Lombardo, P., Benson, W. R., *J. Assoc. Offic. Anal. Chemists* (1970) **53,** 261–2.
(19) Bagley, G. E., Reichel, W. L., Cromartie, E., *J. Assoc. Offic. Anal. Chemists* (1970) **53,** 251–61.
(20) Holden, A. V., Marsden, K., *Nature (London)* (1967) **216,** 1274–6.
(21) Jensen, S., Widmark, G., Institute of Analytical Chemistry, University of Stockholm, Sweden, OECD Preliminary Study 1966–67 (Sept. 14, 1967).
(22) Egan, H. J., *J. Assoc. Offic. Anal. Chemists* (1967) **50,** 74–84.
(23) Robinson, J., Richardson, A., Elgar, K. E., *ACS Natl. Meeting, 152nd, New York, Sept. 11–16, 1966.*
(24) Bowman, M. C., Beroza, M. J., *J. Assoc. Offic. Anal. Chemists* (1965) **48,** 943–52.
(25) Beynon, K. I., Elgar, K. E., *Analyst* (1966) **91,** 143–75.
(26) Goulden, R., Goodwin, E. S., Davies, L., *Analyst* (1963) **88,** 951–8.
(27) Klein, A. K., Watts, J. O., *J. Assoc. Offic. Anal. Chemists* (1964) **47,** 311–6.
(28) Mendoza, C. E., Wales, P. J., McLeod, H. A., McKinley, W. P., *J. Assoc. Offic. Anal. Chemists* (1968) **51,** 1095–101.
(29) Sans, W. W., *J. Agr. Food Chem.* (1967) **15,** 192–8, and personal communication, 1969.
(30) Hammence, J. H., Hall, P. S., Caverley, D. J., *Analyst* (1965) **90,** 649–56.
(31) Pionke, H. B., Chester, G., Armstrong, D. E., *Analyst* (1969) **94,** 900–3.
(32) Chau, A. S. Y., Cochrane, W. P., *Bull. Environ. Contam. Toxicol.* (1970) **5** (No. 2), 133–8.
(33) Hoffman, D., Rathkamp, G., *Beitr. Tabakforschung* (1968) **4,** 201–14.
(34) Fries, G. R., Marrow, G. S., Gordon, C. H., *J. Agr. Food Chem.* (1969) 860–2.
(35) Gunther, F., Blinn, R. C., *J. Agr. Food Chem.* (1957) **5,** 517–9.

(36) Chau, A. S. Y., Cochrane, W. P., *J. Assoc. Offic. Anal. Chemists* (1969) **52**, 1220–6.
(37) Bird, C. W., Cookson, R. C., Crundwell, E., *J. Chem. Soc.* (1961) 4809–16.
(38) Weincke, W. W., Burke, J. A., *J. Assoc. Offic. Anal. Chemists* (1969) **52,** 1277–80.
(39) Skerrett, E. J., Baker, E. A., *Analyst* (1960) 184–7.
(40) Chau, A. S. Y., Cochrane, W. P., *Bull. Environ. Contam. Toxicol.* (1970) **5** (5), 435–9.
(41) Soloway, S. B., U. S. Patent **2,676,131**; C.A. (1954) **48,** 8473e.
(42) Noren, K., *Analyst* (1968) **93,** 39–41.
(43) Osadchuk, M., Wanless, E. B., *J. Assoc. Offic. Anal. Chemists* (1968) **51,** 1264–7.
(44) Cochrane, W. P., Chau, A. S. Y., *J. Assoc. Offic. Anal. Chemists* (1968) **51,** 1267–70.
(45) Cochrane, W. P., Chau, A. S. Y., *Bull. Environ. Contam. Toxicol.* (1970) **5** (No. 3), 251–4.
(46) Bevenue, A., Chee Yee Yeo, *Bull. Environ. Contam. Toxicol.* (1968) **4,** 68–76.
(47) Tu, C. U., Miles, J. R. W., Harris, C. R., *Life Sci.* (1968) **7,** 311–22.
(48) Miles, J. R. W., Tu, C. M., Harris, C. R., *J. Econ. Entomol.* (1969) **62,** 1334–8.
(49) Shell International Research, Belgium Patent **609,983** (May 1963); C.A. (1963) **58,** 1434a.
(50) Singh, J., *Bull. Environ. Contam. Toxicol.* (1969) **4,** 77–9.
(51) Cochrane, W. P., Chau, A. S. Y., *Chem. Ind. (London)* (1968) 1696–7.
(52) Cochrane, W. P., *J. Assoc. Offic. Anal. Chemists* (1969) **52,** 1100–5.
(53) Cochrane, W. P., Forbes, M., Chau, A. S. Y., *J. Assoc. Offic. Anal. Chemists* (1970) **53,** 769–74.
(54) Vollner, L., Klein, W., Korte, F., *Tetrahedron Letters* (1969) **34,** 2967–70.
(55) Chau, A. S. Y., *J. Assoc. Offic. Anal. Chemists* (1969) **52,** 1240–8.
(56) Gorbach, S. G., Christ, O. E., Kellner, H. M., Kloss, G., Borner, E., *J. Agr. Food Chem.* (1968) **16,** 950–3.
(57) Kawano, Y., Bevenue, A., Beckman, H., Erro, F., *J. Assoc. Offic. Anal. Chemists* (1969) **52,** 167–72.
(58) Miller, G. A., Wells, C. E., *J. Assoc. Offic. Anal. Chemists* (1969) **52,** 548–53.
(59) Gunther, F., *Ann. New York Acad. Sci.* (1969) **160,** 72–81.
(60) Bontoyan, W., *J. Assoc. Offic. Anal. Chemists* (1966) **49,** 1169–74.
(61) Crissman, W. I., *Agr. Chem.* (July 1969) 16–18.
(62) Castro, C. E., Kray, W. C., Jr., *J. Am. Chem. Soc.* (1963) 2768–73.
(63) Kray, W. C., Jr., Castro, C. E., *J. Am. Chem. Soc.* (1964) **86,** 4603–8.

Received June 12, 1970.

3

A Review of Enzymatic Techniques Used for Pesticide Residue Analysis

D. C. VILLENEUVE

Food and Drug Directorate, Department of National Health and Welfare, Ottawa, Canada

A review is presented on the use of enzymatic techniques for pesticide residue analysis. Anticholinesterase and anticarboxylesterase procedures, which comprise the majority of such techniques, are described with reference to the theory behind their use, the different methods of assay, their limits of detection, and their advantages and disadvantages. Other enzymatic techniques are also discussed.

The fact that many pesticides inhibit enzymes *in vitro* has led to the introduction of various analytical methods for the detection and estimation of pesticide residues. This paper will not describe in detail all the enzymatic techniques used for pesticide residue analysis but rather will attempt to categorize and describe briefly the main types with reference to the theory behind their use, their practical application, and associated problems. This will be followed by a brief discussion on the advantages and disadvantages of these techniques.

Cholinesterase Techniques

Theory. Cholinesterase inhibition by organophosphorus insecticides has been the subject of several excellent reviews by O'Brien (*1*, *2*) and Heath (*3*). The basis of toxic action of organophosphates and carbamates in mammals is generally associated with their ability to inhibit cholinesterase in the central and peripheral nervous systems where it plays an important role in the transmission of nerve impulses.

Organophosphates and carbamates react with cholinesterase in a manner similar to the reaction of cholinesterase with acetylcholine and can be depicted as follows:

$$\mathrm{EOH} + \mathrm{AX} \rightleftarrows \mathrm{EOH.AX} \rightarrow \mathrm{EOA} \xrightarrow{H_2O} \mathrm{EOH} + \mathrm{A}^- + \mathrm{H}^+ \qquad (1)$$
$$\searrow \mathrm{X}^- + \mathrm{H}^+$$

where EOH represents cholinesterase (OH being a serine OH in the active site), AX is either acetylcholine or an organophosphate or carbamate insecticide, and X is the leaving group—*i.e.*, choline in acetylcholine, *p*-nitrophenol in paraoxon, or 1-naphthol in carbaryl (*2*). The symbol A can designate either the acetyl group, a dialkyl phosphoryl group, or a methyl carbamyl group. Thus, the reaction involves a complex formation followed by phosphorylation, acetylation, or carbamylation of the enzyme, and finally hydrolysis. If the concentration of inhibitor is relatively large (10^{-5} or 10^{-6} *M*) as compared with the concentration of substrate (10^{-10} *M*), the over-all reaction can be considered a first-order bimolecular type and can be described in terms of the following equation:

$$\log P = 2 - \frac{k_i I}{2.3} t \qquad (2)$$

where P is the percent activity of the enzyme remaining after incubation of I concentration of inhibitor for time t, and k_i is the bimolecular rate constant. Thus, for any value of t there is a concentration of inhibitor which would reduce P, the percent activity, to 50%. This value is known as the I_{50}, and because its value is often small, the negative logarithm of the I_{50}, the pI_{50}, is used. Table I gives the I_{50} and pI_{50} of several organophosphorus and carbamate insecticides, using human plasma cholinesterase as enzyme source.

Table I. Cholinesterase-Inhibitive Effects[a] of Several Organophosphate and Carbamate Insecticides

		Quantity to Produce 50% Inhibition[b]	
Insecticide	*Mol. Wt.*	*Molar Concn., I*	*pI*
Coroxon[c]	346.5	3.16×10^{-7}	7.50
DDVP	221.0	6.30×10^{-6}	6.80
DFP	184.2	5.01×10^{-8}	8.70
Guthoxon[d]	301.0	6.31×10^{-5}	5.80
Parathion	291.0	2.43×10^{-5}	5.39
Paraoxon	275.0	1.88×10^{-8}	8.27
Ruelene	291.5	5.62×10^{-4}	4.75
Phosphamidon	299.5	6.7×10^{-7}	7.83
Carbaryl	201.0	1.2×10^{-5}	5.08

[a] Data taken from Giang and Hall (*4*), O'Brien (*5*), and Voss (*6*).
[b] Human plasma cholinesterase used as enzyme source.
[c] Oxygen analog of Co-Ral.
[d] Oxygen analog of Guthion.

From the I_{50} and t, one can calculate, using Equation 2, the k_i value if first-order kinetics are followed. Main (*7*) has developed a kinetic treatment that determines both the affinity of the inhibitor for the enzyme and the ability to phosphorylate the enzyme once binding has occurred. With this treatment, the k_i from Equation 2 cannot be considered as a simple bimolecular rate constant but as a combination of a unimolecular rate constant and an affinity constant. However, the measurement of the k_i as outlined in Equation 2 does provide a satisfactory measure of the inhibitory potency in most cases.

Methods of Assessing Cholinesterase Inhibition. The increased use of cholinesterase-inhibiting insecticides has stimulated research in many areas of scientific endeavor. One such area has been concerned with the *in vivo* toxicological properties of cholinesterase inhibitors. However, the area of concern here is in the field of analytical chemistry, where cholinesterases are used as a tool for the quantitative determination of unknown amounts of inhibitors (*8*). Such procedures are frequently used for the analysis of certain pesticide residues and can be categorized into the following types of methods:

POTENTIOMETRIC OR ΔpH METHODS. One such method, probably the first ever used, was the one published by Giang and Hall (*4*) for the determination of tetraethylpyrophosphate (TEPP) and parathion in plant material. One problem was that parathion was a weak inhibitor *in vitro* and required conversion to its oxygen analog to obtain sufficient sensitivity. This was accomplished by treating parathion with a mixture of concentrated and fuming nitric acids. It required 42.5 μg of parathion to produce 50% inhibition, as compared with 0.031 μg for paraoxon. The method consisted of extracting the pesticide from the plant, incubating an aliquot of the extracted pesticide with a standard bovine cholinesterase solution, determining the remaining cholinesterase activity after incubation, and comparing the percent inhibition with a standard curve in order to obtain the quantity of material. Cholinesterase activity was measured using acetylcholine bromide as substrate and measuring the change in pH caused by the release of acetic acid. The "percent inhibition" (*9*) is calculated as follows:

$$\Delta\text{pH} = \text{pH (initial)} - \text{pH (final)}$$

$$\%\ \text{Inhibition} = 1 - \frac{\Delta\text{pH (sample)}}{\Delta\text{pH (control)}} \times 100$$

TITRIMETRIC METHODS. These methods involve the titration of acetic acid liberated by the hydrolysis of acetylcholine, which is normally accomplished with a "pH-stat" so that a constant pH is maintained. The advantages of these methods over the ΔpH methods are that the measured activity is not directly influenced by the buffer strength of the solution (*10*) and that a constant pH is maintained during the enzyme reaction. This type of method has been used directly for the determination of parathion while other workers used a microtitrimetric method to obtain greater sensitivity (*11*).

MANOMETRIC METHODS. These methods involve the measurement of carbon dioxide liberated from the action of the released acetic acid on sodium bicarbonate. DuBois and Cotter (*in* Ref. *10*) used this type of method to determine Dipterex in milk.

COLORIMETRIC METHODS. *Measurement of Unchanged Acetylcholine.* This colorimetric technique depends on measuring the unchanged acetylcholine with hydroxylamine to produce acetohydroxamic acid which yields a purple color with excess ferric chloride after acidification. Cook (*12*) originally applied this method to pesticide residue analysis on paper chromatograms.

Indophenyl Acetate. A colorimetric method for the analysis of cholinesterase-inhibiting insecticides using indophenyl acetate as substrate was published by Archer and Zweig in 1959 (*13*). The procedure was based on the measurement of the hydrolysis product of indophenyl acetate at 625 μ at a pH of 8.0. Three different enzymatic methods, a potentiometric, a paper chromatographic, and the colorimetric one mentioned, were compared by determining residue levels of carbaryl on peaches. Good agreement was shown by all three methods. Beam and Hankenson (*17*) used the same method to determine known amounts of Sevin, Trithion, parathion, malathion, Guthion, Dipterex, and ronnel in milk. Sensitivity ranged from 0.008 to 0.2 ppm.

Phenol Red. The first publications of an automated procedure for the measurement of cholinesterase inhibitors are those of Winter (*14*) and Winter and Ferrari (*15*). The method employed an Autoanalyzer instrumental system wherein the extracts containing the insecticide were incubated with a standard cholinesterase solution at 37°C. A continuous sample from the incubation bath is buffered and mixed with acetylcholine iodide. After a second incubation, the acetic acid released by the action of the uninhibited cholinesterase is measured colorimetrically, using phenol red as the indicator. More recently, Fischl *et al.* (*16*) reported a method for rapid detection of organic phosphate pesticides in serum. Strips of filter paper were impregnated with a buffered acetylcholine substrate solution containing phenol red as an indicator. When no inhibition is present, the acid released from the action of cholinesterase turns the paper yellow. When cholinesterase has been inhibited, the paper turns pink-to-violet.

Acetylthiocholine Iodide. Another automated procedure for cholinesterase inhibition studies has been used by Levine *et al.* (*18*) and by Voss (*19*). The method uses acetylthiocholine iodide as substrate and dithiobisnitrobenzoic acid (DTNB). Cholinesterase splits the substrate, and the thiocholine released reduces the DTNB to the yellow anion of thionitrobenzoic acid, whose absorbance is measured at 420 μ.

Fluorogenic Substrates. Guilbault and Kramer (*20*) published a method using a fluorometric assay for anticholinesterase compounds. The substrates used were nonfluorescent compounds, the acetyl and butyl esters of 1- and 2-naphthol, which are hydrolyzed by cholinesterase to highly fluorescent materials. The rate of change of fluorescence was related to enzyme activity, and inhibition was measured by decreased rate of change in the production of fluorescence.

A semiquantitative test for the identification of cholinesterase-inhibiting pesticides has been described by Schultzmann and Barthel (*21*). Indoxyl acetate was used as the substrate, in conjuction with a two-dimensional thin-layer chromatographic technique. It consisted of spotting the cleaned-up samples on a silica gel thin-layer plate, developing the plate in the appropriate solvent systems, and spraying the plate with a horse serum cholinesterase solution and indoxyl acetate. The cholinesterase-inhibiting compounds appeared as white spots on a blue background, and sensitivity was in the submicrogram range.

AGAR-AGAR DIFFUSION TECHNIQUES. Benyon and Stoydin (*22*) described a procedure in which a layer of agar-agar 5 mm thick containing cholinesterase and bromthymol blue at a pH of 7.8–7.9 was used. A 2-propanol solution (0.1 ml) of the insecticide was applied to a hole in the agar and was allowed to diffuse for 18 hours at room temperature. The plate was then sprayed with acetylcholine chloride solution, and the blue layer turned yellow within 30 minutes except in those areas where the enzyme was inhibited, which remained as blue circles. The diameter of these circles was proportional to the amount of the inhibitor. As little as 3 ng of parathion could be detected by this technique.

FLUORESCENT EQUILIBRIUM PROBES. Himel and co-workers (*23, 24, 25*) have synthesized active-site-directed fluorescent equilibrium probes which are competitive inhibitors of the active site of cholinesterase enzymes. The fluorescence intensity of the probe–enzyme complex is decreased by any foreign molecule (insecticide) which competes with the equilibrium fluorescent probe for the active site of the enzyme or which changes the equilibrium dynamics by exo area reaction with the enzyme. This highly specific and sensitive spectroscopic method is being developed as an analytical method for insecticides (*26*).

Carboxylesterase Inhibition

Theory. In the Report of the Commission on Enzymes (*27*), carboxylesterases are included in the general group "Hydrolases" (E.C.3.) and more specifically in the category carboxylic ester hydrolases (E.C.3.1.1.1.). These enzymes are responsible for catalyzing the general reaction, carboxylic ester + $H_2O \rightleftharpoons$ alcohol + carboxylic acid. These enzymes have been found in tissues of many species of insects, animals, and plants (*28, 29, 30, 31, 32, 33, 34, 35, 36*). However, the distinction between esterases and other hydrolytic enzymes has become less clear in the light of evidence that certain proteolytic enzymes (trypsin, chymotrypsin, thrombin) can hydrolyze carboxyl esters and are inhibited by substances known to be potent esterase inhibitors (*37, 38*). As with cholinesterases, inhibition involves the reaction of the phosphate with the enzyme to form an acyl derivative (phosphorylated enzyme). The phosphorylated enzyme is quite stable and prevents the action of the esterase on other substrates. The phosphorylation of the enzyme is the rate-limiting step and depends upon the "fit" of the compound on the enzyme and the ability of the compound to phosphorylate a serine or histidine at the enzyme's active site. The detailed mechanism of their inhibition by organophosphorus pesticides has been discussed by several workers and procedures have been worked out for determining rate constants of this step (*7, 39, 40, 41*). It has been suggested (*7*) that the bimolecular rate constant (k_i) is the most reliable criterion to measure the inhibitory power of an organophosphate for an esterase. However, criteria such as percent inhibition and *I* values are considered to be loosely defined functions of the k_i and may be used.

Techniques. In 1962, McKinley and Read (*42*) developed an esterase-inhibition technique for the detection of organophosphate pesticide residues on paper chromatograms. The procedure involved conversion of the thiophosphates with bromine to yield active esterase inhibitors, the inhibition by the pesticide of the esterases from a beef liver homogenate sprayed onto the chromatogram, the hydrolysis of the substrate (α-naphthyl acetate) which was sprayed onto the paper after the liver homogenate had dried, and the development of a background color between Fast Blue RR and the hydrolysis product, α-naphthol.

This technique was adapted to thin layer chromatography by Ackermann (*43*). The substrates used were 1-naphthylacetate, 2-azobenzene-1-naphthylacetate, and indoxylacetate. Sensitivity was in the nanogram range. A later study (*44*) was carried out using different activation techniques in an attempt to make the procedure more sensitive. Mendoza *et al.* (*45*) used indoxyl and substituted indoxyl and naphthyl acetates as substrates. Indigo compounds produced by the hydrolysis of these substrates gave colored backgrounds, leaving the sites of inhibition by pesticides as white spots; the colors produced were blue for indoxyl and 5-bromoindoxyl acetates, turquoise for 5-bromo-4-chloroindoxyl acetate, and pink for 5-bromo-6-chloroindoxyl acetate. Sensitivity of detection was in the nanogram range, and the background and spots were stable for months. Wales and co-workers (*46*) reported a procedure for the semiquantitative determination of some organophosphorus pesticide residues in plant extracts using Mendoza's procedure (*44*). Ackermann (*46*) experimented with different activation techniques to make his thin-layer technique more sensitive. Other workers (*47*) have developed a method to estimate parathion, malathion, and diazinon in lettuce using a carboxylesterase-inhibition assay. The method involved an extraction procedure, followed by sweep codistillation and thin layer chromatography. The pesticides were scraped off the thin layer plate and used in a carboxylesterase assay using *o*-nitrophenyl propionate as substrate. Sensitivity ranged from 1.0 ppm for parathion to 8.0 ppm for malathion and 0.75 ppm for diazinon.

Other Techniques

DDT inhibits carbonic anhydrase and has been made the basis of a quantitative method which can determine as little as 0.2 μg of DDT (*11*). More recently, Guilbault *et al.* (*20*) published a method for the determination of methyl parathion, aldrin, and heptachlor based on the inhibition of acid and alkaline phosphatases by these substances. The substrate was umbelliferone phosphate which was cleaved by the phosphatases to the fluorescent compound umbelliferone. Decreased fluorescence was used as a direct measure of the inhibitor, and the sensitivity was 5 ppm for methyl parathion and aldrin and 50 ppm for heptachlor. Geike (*48*) reported that some organochlorine insecticides will inhibit bovine esterase after exposure to UV irradiation. Ordinarily, these compounds activate liver esterase *in vitro*.

Advantages and Disadvantages of Enzyme Techniques for Residue Analysis

The methods referred to previously are summarized in Table II, along with their reported limits of sensitivity and details on their application to residue analysis.

The most obvious shortcoming of these methods is that very few have been applied to residue analyses in crops or foods. A major problem in the application of these methods to residue analysis is sample extraction and cleanup. It is well known (*8, 49, 51*) that constituents of certain

Table II. Summary of the Different Methods

Method of Assay	*Ref.*	*Pesticides Used*
Potentiometric	(*4*)	Parathion, TEPP, Paraoxon
Titrimetric	(*50*)	Parathion
Manometric	(*52*)	Dipterex
Colorimetric	(*12*)	Systox
	(*13*)	Trithion, Sevin, Guthion, Thimet
	(*17*)	Sevin, Trithion, parathion, malathion, Guthion, ronnel
	(*15*)	Thimet, parathion, diazinon, malathion
	(*53*)	Monocrotophos
	(*6*)	C–8353 (carbamate)
Fluorogenic	(*20*)	Systox, parathion
	(*54*)	DFP, methyl parathion, parathion, dimethyl chlorthiophosphate, diethyl chlorphosphate
	(*55*)	Methyl parathion, aldrin, heptachlor
	(*36*)	29 Organophosphorus and carbamate insecticides
	(*8*)	Birlane, dichlorvos mevinphos
Carboxylesterase	(*42*)	22 Organophosphate insecticides
	(*56*)	32 Organophosphorus and carbamate insecticides
	(*43*)	8 Organophosphorus and carbamate insecticides
	(*44*)	10 Organophosphorus insecticides and Sevin
	(*45*)	7 Organophosphorus insecticides
	(*47*)	Parathion, malathion and diazinon

plants—*i.e.*, tea, potatoes, tobacco—require further cleanup because of enzyme-inhibiting impurities. Many different procedures have been used for extraction and cleanup, and the selection of a suitable method will depend on what crop or food material is being analyzed and the pesticide or pesticides used.

Another disadvantage of enzyme techniques is that they lack the specificity to distinguish individual pesticides when present as a mixture. Most of the methods reported here (Table II) have been developed for single compounds, and thus where more than one pesticide is present, preliminary separation techniques are necessary. Both paper and thin layer chromatographic techniques do not have this disadvantage and

of Analysis Referred to in the Text

Lower Limits of Detection	*Comments*
0.005–5.0 μg/ml	TEPP residues detected on lettuce at levels of 3 ppm and up
1.0 mg/10 cu. m. air	Method applied to the determination of parathion in air
0.75 ppm	Used for the determination of Dipterex in milk
0.60 ppm	Method used for the estimation of systox on apples
0.001–0.1 μg/ml	Sevin determined as a residue on peaches at 35 ppm
0.008–0.3 ppm	Determined singly as residues in milk
0.1–1.0 μg/ml	Procedure applied to standards only
0.02 ppm	Procedure applied to standards only
0.2 ppm	Autoanalyzer procedure applied to apples, cabbage, rice, and wheat
1–10 μg/ml	Procedure applied to standards only
0.4–16 μg/ml	Procedure applied to standards only
0.05–1.0 μg/ml	Procedure applied to standards
0.5–10 μg/ml	Thin layer technique applied to standards only. Various methods of sample extraction and cleanup described
0.01–0.3 μg/ml	Mevinphos determined in crop extracts at 0.05 ppm
0.5–5.0 μg	Paper chromatographic technique applied to standards only
0.01–3 μg	Paper chromatographic method applied to extracts of lettuce, strawberries, and apples
0.2–10 ng	TLC procedure; applied to standards only
0.001–0.1 μg	TLC procedure; applied to standards only
0.2–4.0 ppm	TLC procedures; applied to potato extracts
0.75–8.0 ppm	Combined TLC–enzyme-inhibition procedure. Insecticides determined simultaneously in lettuce extracts

serve as excellent screening techniques. However, before the other enzymatic methods can be successfully applied to the determination of multi-pesticide residues, a preliminary separation of the specific pesticides is essential.

A third problem is that many organophosphorus pesticides are poor esterase inhibitors *in vitro* and need to be converted to their oxygen analogs in order to obtain sufficient inhibitory potency. Some of the procedures used for this conversion are shown in Table III. Again, the choice of a suitable procedure will depend on the type of assay used and the pesticides in question. In our laboratories, bromine vapor is used

Table III. Methods for Converting Organophosphorus Pesticides to Active Esterase Inhibitors

Method	*Reference*
Cold fuming nitric acid	(*4*)
Dilute bromine water	(*9*)
N–Bromosuccinamide	(*9, 21*)
UV light	(*57*)
H_2O_2–Acetic acid	(*17*)
Bromine vapor	(*47*)
Peracetic acid	(*37*)
m–Chloroperbenzoic acid	(*37*)
H_2O_2	(*37*)

to accomplish activation of organophosphorus pesticides after separation on thin layer chromatograms. However, the bromine also converts some plant constituents into antiesterase compounds and thus adds a new source of interference.

There are other disadvantages, some of which are peculiar to the method employed. For example, in the case of the automated analysis described by Voss (*8*), some of the pump tubing used was susceptible to contamination by certain insecticides.

One of the most important advantages of these methods is their sensitivity. Generally, these methods measure submicrogram quantities of the insecticide in question and are more sensitive than most chemical methods. Moreover, enzymatic methods can detect insecticides that are converted into metabolites with a high inhibitory potency. Enzymatic methods can be simple and rapid. Automated analyses, for example, provide simple high-precision techniques with short incubation periods, high sensitivity, and adaptability to routine analyses. As mentioned previously, such methods are invaluable as screening techniques.

Summary

A description of several anticholinesterase and anticarboxylesterase techniques as they are used in pesticide residue analysis has been given. Some other enzyme techniques have also been mentioned, as well as the advantages and disadvantages of these techniques.

Literature Cited

(1) O'Brien, R. D., "Toxic Phosphorus Esters," Ch. 3, Academic, New York, 1960.
(2) O'Brien, R. D., "Insecticides: Action and Metabolism," pp. 39–54, 86–95, Academic, New York, 1967.
(3) Heath, D. F., "Organophosphorus Poisons," Introductory Chapter, Pergamon, Oxford, 1961.

(4) Giang, P. A., Hall, S. A., *Anal. Chem.* (1951) **23,** 1830–4.
(5) O'Brien, R. D., *J. Agr. Food Chem.* (1963) **11,** 163.
(6) Voss, G., *Bull. Environ. Contam. Toxicol.* (1968) **3,** 343.
(7) Main, A. R., *Science* (1964) **144,** 992–3.
(8) Voss, G., *Residue Rev.* (1968) **23,** 71–95.
(9) Archer, T. E., "Enzymatic Methods" *in* "Analytical Methods for Pesticides, Plant Growth Regulators and Food Additives," Vol. 1, Ch. 14, G. Zweig, Ed., Academic, New York, 1963.
(10) Gage, J. C., *Advan. Pest Control Res.* (1961) **4,** 183–210.
(11) Schechter, M. S., Hornstein, I., *Advan. Pest Control Res.* (1957) **1,** 353–447.
(12) Cook, J. W., *J. Assoc. Offic. Anal. Chemists* (1954) **37,** 561–4.
(13) Archer, T. E., Zweig, G., *J. Agr. Food Chem.* (1959) **7,** 178–81.
(14) Winter, G. D., *Ann. N. Y. Acad. Sci.* (1960) **87,** 875.
(15) Winter, G. D., Ferrari, A., *Residue Rev.* (1964) **5,** 139.
(16) Fischl, J., Pinto, N., Gordon, C., *Clin. Chem.* (1968) **14,** 371–3.
(17) Beam, J. E., Hankenson, D. J., *J. Dairy Sci.* (1964) XLVII, 1297–1305.
(18) Levine, J. B., Scheidt, R. A., Nelson, V. A., Technicon Symposium, "Automation in Analytical Chemistry," p. 582, New York, 1965.
(19) Voss, G., *J. Econ. Entomol.* (1966) **59,** 1288.
(20) Guilbault, G. G., Kramer, D. N., *Anal. Chem.* (1965) **37,** 1675–80.
(21) Schultzmann, R. L., Barthel, W. F., *J. Assoc. Offic. Anal. Chemists* (1969) **52,** 151–6.
(22) Benyon, K. I., Stoydin, G., *Nature* (1965) **208,** 748–50.
(23) Himel, C. M., Mayer, R. T., *J. Georgia Entomol. Soc.* (1970) **5,** 31–8.
(24) Himel, C. M., Mayer, R. T., Cook, L. L., *J. Polymer Sci.* (1970) Pt. A-1, **8,** 2219–30.
(25) Himel, C. M., Aboul-Saad, W. G., Uk, S., *J. Agr. Food Chem.*, in press, 1971.
(26) Himel, C. M., private communication, 1970.
(27) Dixon, M., Webb, E. C., "Enzymes," p. 732, Longmans, Green and Co. Ltd., 1965.
(28) Barron, K. D., Bernsohn, J. I., Hess, A., *J. Histochem. Cytochem.* (1961) **9,** 656–60.
(29) Bernsohn, J., Barron, K. D., Hess, A., *Proc. Soc. Exptl. Biol. Med.* (1961) **108,** 71–3.
(30) Cook, J. W., Blake, J., Yip, G., Williams, M., *J. Assoc. Offic. Agr. Chemists* (1958) **41,** 399–407.
(31) Ecobichon, D. J., Kalow, W., *Can. J. Biochem. Physiol.* (1961) **39,** 1329–32.
(32) Holmes, R. S., Masters, C. J., *Biochem. Biophys. Acta* (1968) **151,** 147–58.
(33) Lawrence, S. H., Melnick, P. J., Weimer, H. E., *Proc. Soc. Exptl. Biol. Med.* (1960) **105,** 572–5.
(34) Main, A. R., Braid, P. E., *Biochem. J.* (1962) **84,** 255–63.
(35) Norgaard, M. J., Montgomery, M. W., *Biochim. Biophys. Acta* (1968) **151,** 587–96.
(36) Seume, F. W., Casida, J. E., O'Brien, R. D., *J. Agr. Food Chem.* (1960) **8,** 43–7.
(37) Dixon, G. H., Neurath, H., Pechere, J. F., *Ann. Rev. Biochem.* (1958) **27,** 489–532.
(38) Fruton, J. S., *Harvey Lecture Ser.* (1955–56) **51,** 64–87.
(39) Aldridge, W. N., *Biochem. J.* (1950) **46,** 451–60.
(40) Main, A. R., Dauterman, W. C., *Nature* (1963) **196,** 551–3.
(41) Main, A. R., Iverson, F., *Biochem. J.* (1966) **100,** 525–31.
(42) McKinley, W. P., Read, S. I., *J. Assoc. Offic. Agr. Chemists* (1962) **45,** 467–73.

(43) Ackermann, H., *Nahrung* (1966) **10,** 273–4.
(44) Ackermann, H., *J. Chromatog.* (1968) **36,** 309–17.
(45) Mendoza, C. E., Wales, P. J., McLeod, H. A., McKinley, W. P., *Analyst* (1968) **93,** 34–8.
(46) Wales, P. J., Mendoza, C. E., McLeod, H. A., McKinley, W. P., *Analyst* (1968) **93,** 691–3.
(47) Villeneuve, D. C., Butterfield, A. G., McCully, K. A., *Bull. Environ. Contam. Toxicol.* (1969) **4,** 232–9.
(48) Geike, F., *J. Chromatog.* (1969) **44,** 95–102.
(49) Ackermann, H., *Nahrung* (1968) **12,** 357–62.
(50) Brown, H. V., Bush, A. F., *Arch. Ind. Hyg.* (1950) **1,** 633–6.
(51) Menn, J. J., McBain, J. B., Dennis, M. J., *Nature* (1964) **202,** 697–8.
(52) DuBois, K. P., Gladys, J. C., *Arch. Ind. Health* (1955) **11,** 53–60.
(53) Voss, Gunther, *J. Assoc. Offic. Anal. Chemists* (1969) **52,** 1027–34.
(54) Guilbault, G. G., Lubrano, G. J., *Anal. Chim. Acta* (1968) **43,** 253–61.
(55) Guilbault, G. C., Sadar, M. H., Zimmer, M., *Anal. Chim. Acta* (1969) **44,** 361–7.
(56) McKinley, W. P., Johal, P. S., *J. Assoc. Offic. Agr. Chemists* (1963) **46,** 840–2.
(57) Ackermann, H., *Arch. Toxikol.* (1969) **24,** 325–31.

RECEIVED June 12, 1970.

4

Flame Detectors for Residue Analysis by GLC

WALTER A. AUE

University of Missouri, Columbia, Mo. 65201

This presentation reviews characteristics of three gas chromatographic detectors: The flame ionization, the flame photometric, and the alkali-flame detector. Modes of operation, sensitivity and selectivity limits, and the relative advantages and disadvantages of these detectors are covered from the viewpoint of the residue chemist. In a final section dealing with recent research of the author's own group, the magnitude and range of the negative alkali flame detector response (inverted peaks) is characterized. It is favored by a smooth and clean alkali salt surface, a large bead bore, and high hydrogen and low carrier gas flows. The phenomenon has been used for qualitative and semiquantitative structure analysis by GLC.

The topic is by no means exhaustively covered by this paper. It would have been a formidable—and redundant—task to do so in light of several excellent monographs and the biannual ACS reviews (*e.g.*, Ref. *1*) in this field. Rather, I have tried to assume the viewpoint of a residue chemist interested in the relevant characteristics of flame detectors, their advantages and disadvantages, their typical application, and, last but not least, some new developments in the area.

For reasons of expediency, the figures are taken exclusively from those at my disposal—*i.e.*, from our own laboratory. No doubt several of the original articles found in the literature—and I have tried to cite most of them—contain better illustrations of the detector characteristics discussed.

Flames in GLC Detectors

Flames are ubiquitous in gas chromatography. They vary in size, energy, and chemical composition. Although high-energy flames are occasionally used to determine gas chromatographic effluents—*e.g.*, as a silicon-specific emission detector (*2*)—the typical flame in this field is a very small hydrogen–air diffusion flame.

Some flames are doped ("sensitized") with various metals, others are plain. These metals can be constantly present in the flame (alkali) or be temporarily present in conjunction with a suitable chromatographic peak [copper (*3, 4*), indium (*5*)]. Several flame variables can be monitored, such as the conductivity, the light emission (or absorption), or the temperature.

In this paper, only three kinds of detectors are discussed in detail. These are the flame ionization detector (FID), the flame photometric detector (FPD), and the alkali flame detector (AFD). All three are highly sensitive systems which have been used extensively in trace analyses. The FID and the FPD represent the plain flame type, in which the flame conductivity (FID) or a selected light emission (FPD) are monitored. The AFD is basically an FID doped with alkali, in which usually the conductivity, occasionally the emission (*6*), generate the detector signal.

Although this report—and the typical residue laboratory—are concentrating on those three types of flame detectors, other devices have demonstrated great potential. Thus, for example, the indium detector first described by Gilbert (*7*) was adopted for GC effluent analysis by Gutsche *et al.* (*5*). Indium chloride or indium bromide (*8*) is formed from volatile halides and determined by its molecular bands in the hydrogen flame. Thus, a highly selective chemical reaction and a spectroscopic determination can combine to a detector system of high specificity.

Flames certainly represent one of the easiest and most efficient ways to build detectors. Therefore, it may come as a surprise to some that flames are not really necessary. This is to say, they are not indispensable —they can be replaced by another source of energy without abandoning the basic detector function.

The FID, for instance, produces the signal from chemionization in the oxidation of combustible carbon radicals, schematically, $CH + O \rightarrow CHO^{+} + e^{-}$. This process can be duplicated on a glowing platinum wire in contact with column effluent and oxygen. It is obvious that differences will be found in the relative response to different types of organic compounds by these two detector versions. Mechanistically, however, they may well be similar. The most recent paper on this subject was presented by Umstead in this symposium (*9*).

At least some of the properties of the alkali flame detector (AFD) can be duplicated in a flameless system. Highly increased effluent conductivity results from the contact between phosphorus compounds and cesium bromide vapor in the gas phase (*10*). On the other hand, alkali-doped platinum electrodes respond strongly to halogen-containing vapors in the so-called "leak detector" (*11*).

A flameless form of the flame photometric detector has, to my knowledge, not been devised. Yet one may venture to predict its eventual arrival. After all, measurement and interpretation of the HPO band spectrum (which is responsible for the FPD's response) were first accomplished from a discharge, not a flame.

Why then the predominance of flames? Small hydrogen diffusion flames can be easily produced, maintained, and regulated. They are not easily disturbed, extinguished, or contaminated under normal conditions and provide energy sources of low noise with often extremely wide linear ranges, with the exception of the flame photometric detector in the sulfur mode, which varies with the square root of concentration. Detectors constructed with flames are fast and usually have a negligible dead volume. It is some of these characteristics which distinguish flames from other energy sources such as electric or electromagnetic discharges.

Besides speed and linear range, the three detectors under consideration can be well characterized by the three classical standards: sensitivity, selectivity, and reliability. These standards will also offer a basis of comparison with some of the other methods of detection presented in this symposium.

It is surprising, I think, how many specific and highly sensitive GLC detectors are at the disposal of residue chemists or have, in fact, been developed specifically for residue analysis. If I were to guess at the historical origin of the three detectors discussed, the first one seems to have resulted from a deliberate effort to construct a GLC detector. The second one may have originated from the ingenious grasping of one of nature's hints which fortunately happened to drop in the right place. And if one considers the close chemical and historical ties between nerve gases and thiophosphate pesticides, one may assume that the third detector, at least in budgetary terms, perhaps represents a sword turned into a plowshare.

The Flame Ionization Detector

The FID is perhaps the most valuable general GLC detector by popular acclaim. Its function and characteristics have been investigated in detail and are described in most books on GLC (*e.g.*, Ref. *12, 13*). A recent paper evaluated its performance in several commercial instruments (*14*). The FID detects virtually all combustible carbon compounds with high sensitivity and fairly predictable relative response. This lack of selectivity is a virtue in some fields—petroleum chemistry, for example—but it has often cast the FID into the role of a standby in the realm of residue analysis. There the FID is employed for cases where a selective detector cannot be used or where knowledge of the amount and variety of background materials is desired.

Put to work on a relatively clean sample, the FID shows impressive credentials. In laboratory routine, one should expect a linear range of 5 to 6 orders of magnitude, minimum detectable amounts of approximately 1 ng, and a very high degree of reliability. With a little bit of cleaning now and then, a good hydrogen flame detector will stay at a relatively constant level of sensitivity longer than the whole instrument may last. It is, of course, ideal for temperature programming or the use of capillary columns.

For residue analysis, the requirements placed upon the FID are more stringent than for other types of GLC analysis. Cleanliness is the mother of good response, of course, but other parameters may be even more important.

Out of a number of authors, Gill and Hartmann (*15*) and more recently Knapp (*16*) discussed possible and advantageous detector and amplifier circuitry. From these studies and general experiences with FID's, a few rules evolved. The battery supplying the necessary potential across the flame should not be situated between the collector electrode and the electrometer. Electric leaks can lead to serious noise problems because of the extremely low currents involved. Two other factors which contribute more to the generation of noise than is commonly realized are an inadequate ground and the presence of vibrations on the cables transmitting the signals. These cables should be well shielded and devoid of any mechanical disturbance.

Various FID modifications have been developed to protect the flame from draft, obtain an optimal air flow pattern, curtail thermionic emission of electrons from the detector jet tip, etc. An interesting development is the use of a "horizontal" flame in a high-efficiency FID designed to minimize combination reactions (*17*).

When first put into use—and every few months thereafter—the flow conditions for optimum response of the FID should be determined. This can be done by the time-honored method of repeated injections while varying the flow rate of air and especially of hydrogen or by a faster method recently publicized (*18*). This test takes but a few minutes to execute but can improve analytical results considerably. To even mention FID optimization may well be redundant. It has been my experience, however, that most gas chromatographs equipped with flame ionization detectors are run under less than ideal flow conditions. In trace analysis, this oversight may be crucial.

Flame ionization detectors can be used advantageously in residue analysis when relatively clean extracts are available, when temperature programming over a wide range is called for, when the residues of interest contain no hetero-atoms or other features which would allow the use of selective detectors, or when other substances present in the injection mixture would disturb a selective detector. To illustrate the latter point with a frustration of our own: The attempted determination of silylated picloram (O-trimethylsilyl-4-amino-3,5,6-trichloropicolinic acid) with the Ni-63 electron capture detector was impossible because the otherwise excellent reagent bis(trimethylsilyl)trifluoroacetamide knocked out the detector for the next half hour following injection (*19*).

The overwhelming majority of pesticides contain one or more hetero-elements. This is true of many pesticides and of almost all compounds which pose serious residue problems. It is this fact that has led to the fast rise and extreme importance of selective detectors in pesticide analysis. The next detector I would like to discuss is, from the strict constructionist point of view, only a slight modification of the old flame. Yet it is a much more temperamental device, highly selective and sensitive for phosphorus-containing compounds, and full of surprises for the chromatographer who attempts to alter one of its features.

The Alkali Flame Detector

The alkali flame detector had a short but turbulent history (*20*). Just how short is a matter of some controversy. The question, in a nutshell, is whether the so-called leak detector (*21*) and the so-called thermionic detector owe their function to the same mechanism. The choice of the word "alkali-flame" in this paper is simply one of convenience. Giuffrida termed her discovery a "thermionic detector" (*22*). It has also been called "Karmen–Giuffrida detector," "phosphorus detector," and "alkali-sensitized flame detector." The term "thermionic," I think, carries mechanistic overtones, and there is little definite knowledge on the mechanism of this detector. "Phosphorus detector" describes its most prominent feature but implies too narrow a range of application. "Alkali flame," on the other hand, is a broad, innocuous, and unobtrusive term. There exists, incidentally, one other detector which uses an alkali flame. It monitors changes in alkali emission and flame conductivity (*6*).

The mechanism of the AFD is disputed. Moesta and Schuff (*23*) assumed the presence of an alkali–oxygen complex on heated platinum which dissociates in the presence of halides. Giuffrida and Ives thought the blue color (caused by phosphorus entering the flame) indicative of a high energy state which could be utilized in the presence of heated alkali (*22*). Karmen, in his latest paper, ascribed the sensitivity of the AFD to phosphorus to increased ionization of the alkali metal in the flame and its sensitivity toward halogen to an increase in volatility of the alkali salt

(*24, 25*). The idea of two different mechanisms for halogen and phosphorus is also supported by Nowak (*26*) and Janák (*27*). Saturno and Cook suggested reactions in the gas phase which led to an ionization of alkali metal atoms (*28*). A similar view is held by Janák (*27*). Page and Woolley assumed an ion-producing reaction in the gas phase to be coupled with the hydrogen (atom/molecule) equilibrium, or rather disequilibrium (*29*). Brazhnikov *et al.* believe that the blue light produced by phosphorus compounds photo-evaporates alkali salt (*20, 30*). A detailed analysis of the experimental data and theoretical assumptions surrounding the various proposed mechanisms would be interesting but beyond the scope of these short remarks which are directed primarily toward the residue chemist.

What distinguishes one alkali flame detector from the other is largely the way in which alkali is brought into the flame. The fact that the performance of the detector depends to a great extent on the physical and chemical structure of the alkali source may serve to explain many seemingly contradictory results. The recent, quite elaborate, review by Brazhnikov *et al.* (*20*) provides a good source of reference for different detector constructions.

There are several commercial companies producing alkali flame detectors. In one version, two flames are stacked, one above the other. The lower, plain hydrogen–air flame burns the sample; the combustion products are swept into the second flame which is doped with a sodium salt deposited on an electrically heated wire (*31*). The upper detector functions as alkali flame detector. Another modification uses a detector jet tip formed from fused salt; the flame burns in contact with the salt surface (*15, 32, 33, 34, 35*). A third form uses an alkali-doped porous metal (*36*) or a platinum capillary filled with potassium hydroxide and carbon (*37*). When the capillary is heated to 900°–1000°, the carbon causes the grain boundaries in the platinum to become enlarged, allowing alkali to diffuse slowly through it. The above remarks should be considered illustrative of detectors on the market and by no means comprehensive. Commercial firms, of course, must add the limitations of the patent situation to the difficulties encountered in constructing these detectors. In fact, some detectors seem to be constructed more for the patent lawyer than for the analyst (*38*).

No such limitations bind the residue chemist who wants to make his own. There are, in my opinion, two easy ways to produce a good alkali flame detector from a suitable commercial FID. Both are inexpensive but may take a little practice. Method number 1 was described by Giuffrida and Ives (*39, 40*). A 26-gauge platinum–iridium wire helix, coated with potassium chloride, is mounted on a detector jet tip such that it is in contact with the flame (*41*). Method number 2 stems from an idea by Coahran, who constructed a sodium sulfate reservoir at the bottom of the flame (*42*). This idea evolved into pressed beads or pellets (*22, 32, 33, 34*), also called salt tips. The performance of various alkali salts has been given considerable attention (*20*); recent studies on tips made from different salts are those by Dressler and Janák (*43*) and by Ebing (*44*).

Giuffrida pointed out that alkali halides suppress the response of halogen compounds in the AFD, but leave the phosphorus response unimpaired. She suggested potassium chloride as the best alkali salt for coating the helix. Fortunately, alkali halides are easily pressed into pellets of considerable mechanical stability, as any IR analyst can attest. A thick IR pellet made from potassium chloride can be drilled to slide over the detector jet tip such that the flame burns in contact with its upper surface. We have used such a simple device with satisfying results for determining phosphate derivatives of hydroxyl and amino compounds obtaining minimum limits in the 1–5 pg range (*45*). Pellets, of course, become somewhat more difficult to produce from sulfates or from salts other than halides.

Once a flame ionization detector has been converted to an alkali flame detector by the addition either of a coated spiral or a pellet, the residue chemist should bear in mind that each detector is a little different from the other and some tinkering with the flow rate and the electrode height (where possible) may be of great benefit. Some alkali flame detectors need an initial conditioning. Some types of pellets, including a commercially available one, are quite susceptible to contamination and the alkali surface has to be cleaned from time to time. The alkali salt itself, of course, should be a high-purity compound.

As amply documented, the most important of the flow rates is that of hydrogen. The alkali flame detector response for phosphorus increases with increasing hydrogen flow. A higher hydrogen flow, of course, produces a bigger flame of higher temperature which is in contact with a

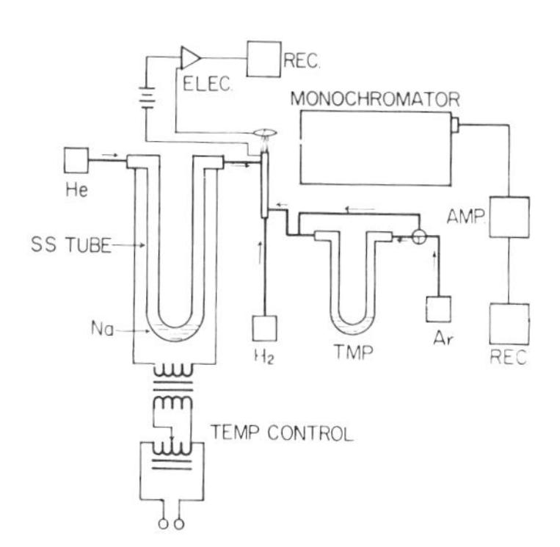

Figure 1. Simultaneous measurement of alkali flame emission and conductivity

He, H_2, Ar: gas supplies with rotameters. Na: liquid sodium or potassium. TMP: Trimethylphosphate, bromobenzene, or iodobenzene.

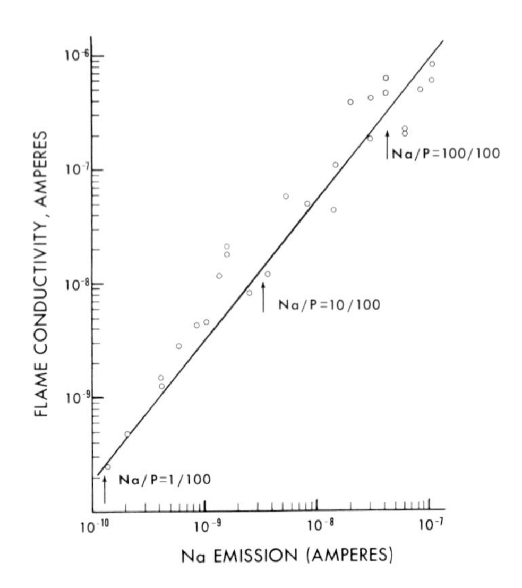

Figure 2. Sodium–trimethylphosphate interaction

P concentration and all flows are kept constant; Na is increased and again decreased by varying the temperature. Na/P atomic ratios are estimated from temperature and flow conditions.

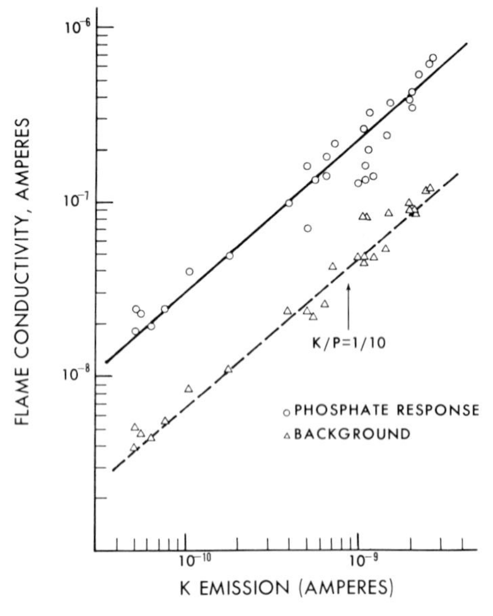

Figure 3. Potassium–trimethylphosphate interaction; experiment as in Figure 2, but with different flow rates

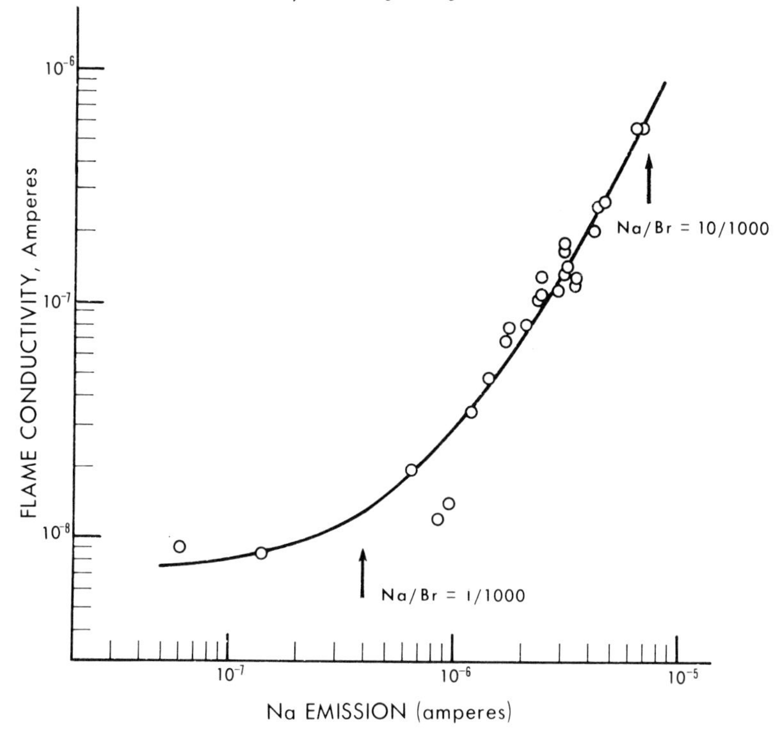

Figure 4. Sodium–bromobenzene interaction

greater area of alkali surface. Consequently, the concentration of alkali in the flame and the background current rises.

It had been of interest to us some time ago whether a flame in which only the alkali content was changed would change its response to phosphorus. Alkali vapor can be brought into the flame in a helium stream at various concentrations by bubbling the helium through liquid alkali metal held at various temperatures. The experimental set-up is shown in Figure 1. The amount of alkali can be calculated from the flow rate of helium and the temperature of the alkali metal; it is monitored by a spectrometer. Constant concentrations of trimethyl phosphate are introduced in a nitrogen stream. Some results obtained with this set-up are shown in Figures 2 and 3 for phosphorus interacting with sodium and potassium and in Figures 4 and 5 for bromine and iodine interacting with sodium (*46*). As a detector, incidentally, this device is decidedly inferior to the pellet or spiral versions (compare also Ref. *26*).

Leaving some obvious mechanistic deliberations aside, the data make clear that the amount of alkali in the flame is influencing, if not determining, the response and that therefore higher hydrogen flow will increase the response. The limit to excessive hydrogen flows used with "regular"

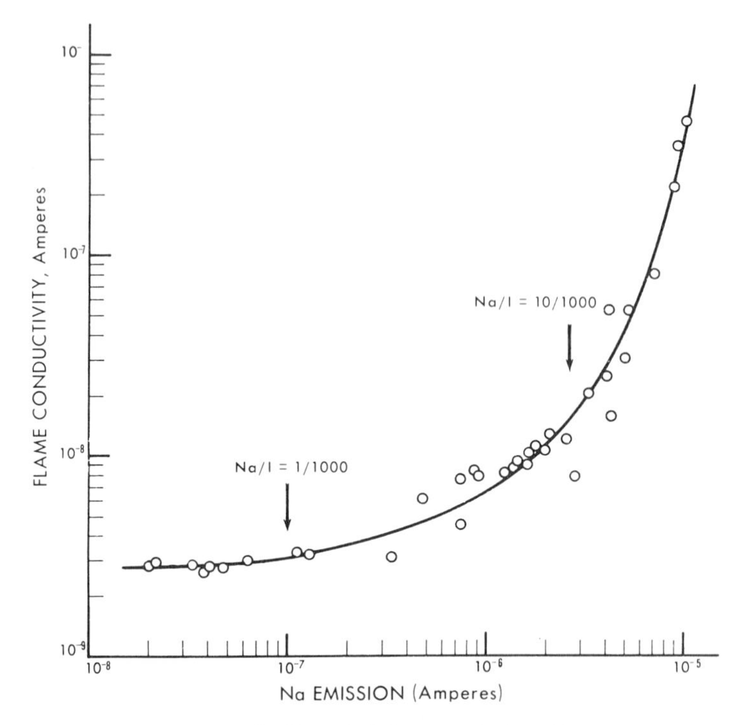

Figure 5. Sodium–iodobenzene interaction

alkali flame detectors are imposed by either the strong background current exceeding the buck-up capability or the linear range of the electrometer, or by a decrease in the effectiveness of the detector as measured by its signal-to-noise ratio (*47*). Noise, unfortunately, also increases strongly with an increase in hydrogen flow.

In detectors which use a single platinum loop as collector electrode, the position of the electrode in regard to the flame is of some importance. If such an electrode is displaced vertically, moving up and down so to speak, it will encounter maxima for different heteroelements at different heights. The resulting response profiles are an interesting field to study, but are mentioned here only to indicate the importance of optimizing the electrode position for residue analysis. Response profiles vary greatly with detector dimensions and flow rates. One example for response maxima of three elements from group five of the periodic system (*48*) is shown in Figure 6. The response is plotted on arbitrary scales and the numbers at the apices indicate the enhancement as compared with a run-of-the-mill hydrogen flame detector. Such a picture, I believe, clearly points out the advantages of electrode height optimization. Depending on the flame shape and the diameter of the electrode, these response profiles can become more or less pronounced.

Since the background current is a measure for the concentration of alkali in the flame, it can also give a rough estimate of how good a detector performance you may rightfully expect. Alkali flame detectors are prone to sensitivity shifts, and in some residue laboratories the background current is checked routinely in the morning before any analysis is performed. Needless to mention that the alkali flame detector needs a closer control of the flow rate, especially that of hydrogen, than the FID. Differential flow controllers, flow-restricting capillaries, and similar devices can greatly improve the AFD's performance (*49*). Carrier gas regulation becomes a problem of the first order when temperature programming is attempted. Contrary to a commonly held belief, temperature programming with alkali flame detectors is possible, as Bostwick and Giuffrida have first shown (*49*). Differential flow controllers and especially a low carrier gas flow can do the trick. An example of a temperature-programmed chromatography (*50*) is shown in Figure 7. The compounds are phosphated amino acid methyl esters, not pesticides, but

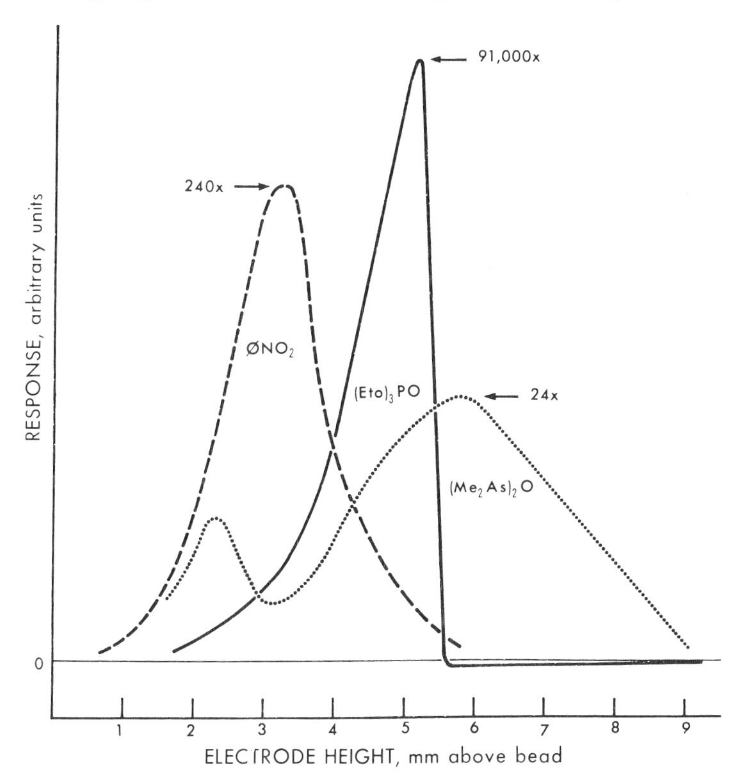

Figure 6. Response profiles measured at extremely high hydrogen flow; Rb_2SO_4-coated ceramic bead in a modified F & M FID; component concentrations are adjusted to give response of comparable size. Figure at apex: enhancement compared with regular FID.

should illustrate my point just as well. For routine pesticide analysis, however, it is highly recommended to work at isothermal conditions whenever possible. A recent example of temperature programming involving nitrogen compounds was published by Hartmann (35).

What type of performance can—or rather should—the residue chemist expect of the alkali flame detector? Ten picograms of parathion should

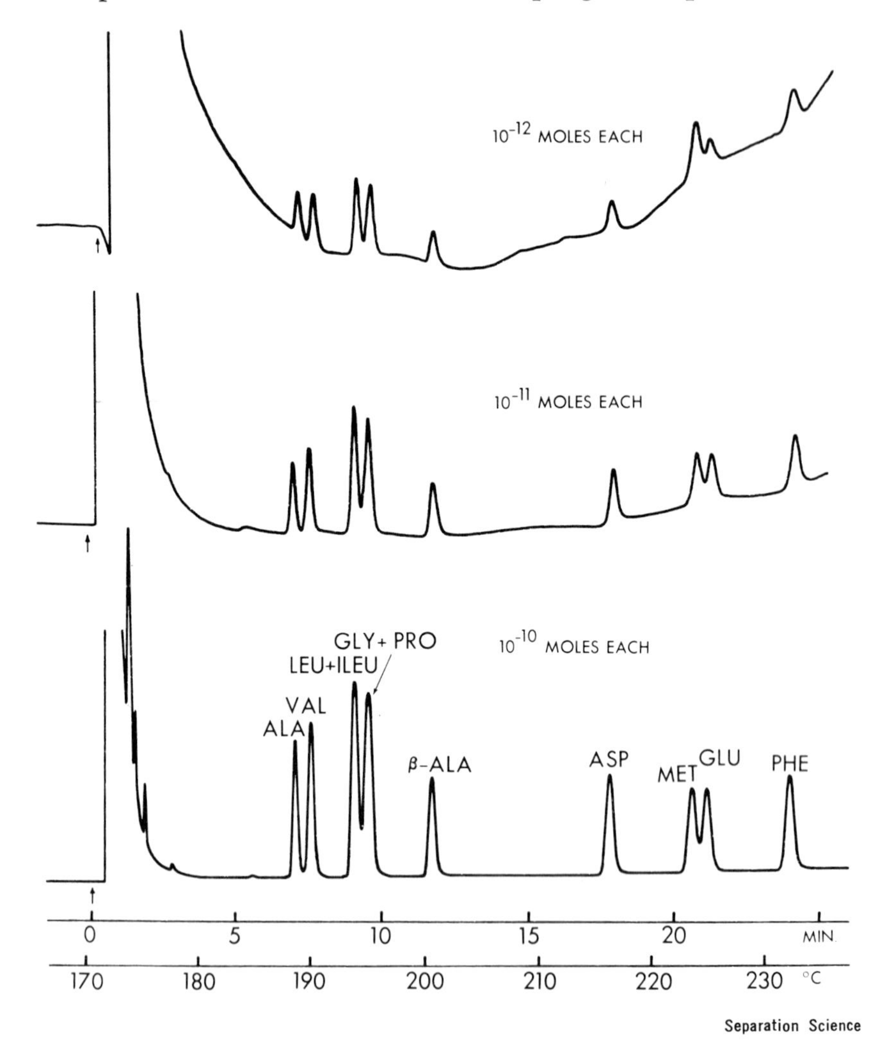

Figure 7. Temperature-programmed GLC of N-diethylphosphate amino acid methyl ester derivatives [$(EtO)_2$ PO-NH-CH(R)-COOMe]. Rb_2SO_4-coated ceramic bead in a Barber-Colman FID. 5% Carbowax 20M on Chromosorb W, AW-DMCS, 60/80 mesh in a 2-m by 2.5-mm i.d. Pyrex U-tube. Flow rates, ml/min.: N_2 16, H_2 40, Air 200.

Reprinted from Ref. 50, p. 687, by courtesy of Marcel Dekker, Inc.

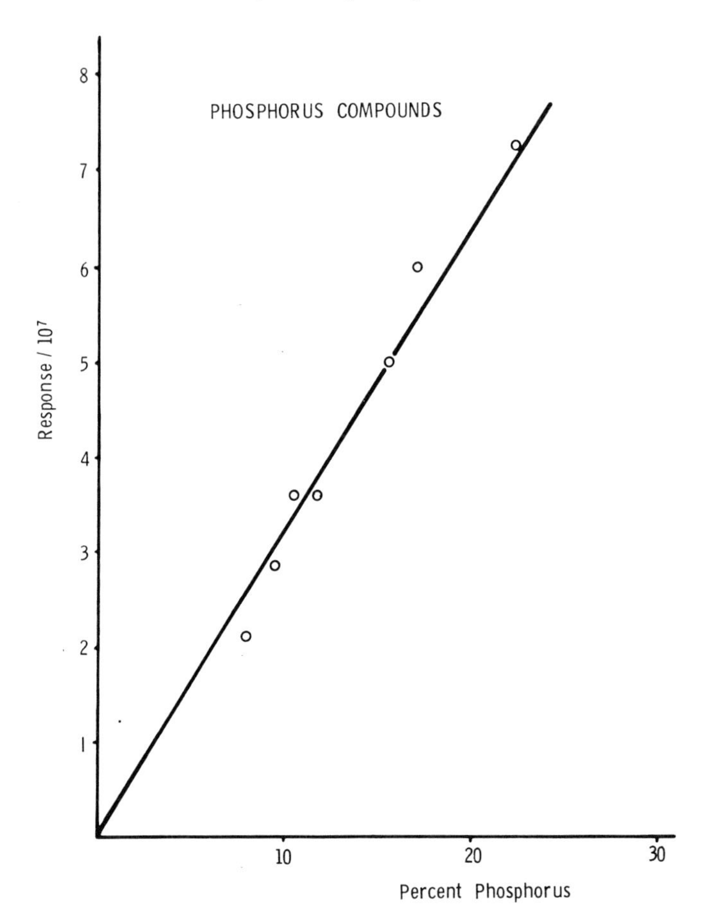

Figure 8. Response of equal weights of phosphorus compounds

Pressed and drilled Rb_2SO_4 pellet in a lab-made AFD (Figure 11). Pellet bore: 1 mm. Electrode: 7-mm i.d., set at 10 mm above the pellet. Flow rates, ml/min.: H_2 35, N_2 50, Air 215.

be clearly detectable. At optimum conditions, one picogram (*51*) and, more recently, ten femtograms (*47*) of parathion have been detected. When the detector is set for halogen or nitrogen response—requiring somewhat different flow conditions and a different alkali source, *e.g.*, rubidium sulfate—1 ng of a typical halogen- or nitrogen-containing compound should be detectable. At optimum conditions, Hartmann reported a minimum detectable limit of 20 pg for azobenzene (*35*), far better than the 500 pg we were able to do on *s*-triazines (*52*). A linear range of three orders of magnitude should be satisfactory, although considerably higher values have been reported (for a discussion, *see* Ref. *20*, p. 24). The de-

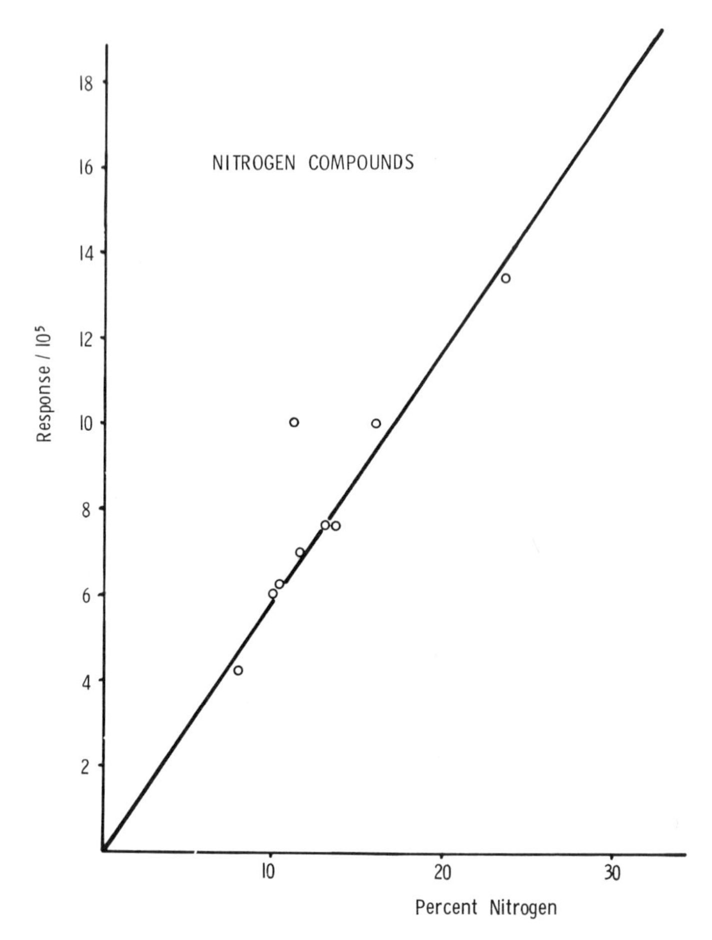

Figure 9. Response of equal weights of nitrogen compounds. Conditions as given in Figure 8, except electrode height 7 mm.

tector response to As (*53, 54*) and S (*55*) compounds is considerably smaller than to N or Cl.

The response from the AFD is roughly proportional to the amount of phosphorus entering the flame. The proportionality of its response to phosphorus and other active elements has been mentioned in several reports (*53, 56, 57, 58*) although some authors do not quite agree (*27, 35*). Figure 8 shows the response of several phosphorus compounds injected in equal weight amounts. Figure 9 shows the same plot for nitrogen-containing compounds.

The predominant use at the present time for the alkali flame detector is undoubtedly the analysis of phosphorus-containing materials. Although halides, nitrogenous compounds, and even an arsenical have been determined with the alkali flame detector—and our group found a lot of fun and a lot of frustration in working in this field—I would be loath to recommend it for routine use in view of such devices as the electrolytic conductivity detector.

Not only is the sensitivity of the AFD much higher for phosphorus than for all the other elements, but also the selectivity is much better. While the selectivity of phosphorus compounds as compared with carbon compounds may be in the order of 10^4 to 10^5, the selectivity for halides is between 10^2 and 10^3, nitrogen occupies the same range, and sulfur and arsenic are more than one order of magnitude lower.

Some time ago, we attempted to illustrate the selectivity of the AFD for P in a somewhat unusual way. We produced a mixture of diethylphosphate derivatives of several typical pesticide hydrolysis products and determined them by alkali flame. We then chose hydrocarbon peaks which we knew would interfere with the phosphorus-containing compounds. Figure 10 shows an experiment in which these hydrolysis products together with a 10,000-fold excess of the hydrocarbons were derivatized with diethylchlorophosphate and the resulting mixture injected into the chromatograph. On close examination of the chromatogram, it becomes clear that we observe a concentration ratio in which the main peaks are the phosphorus-containing derivatives, but some interference from the hydrocarbons begins to show up. Selectivity depends primarily on the amount of alkali in the flame. An increase in hydrogen flow will, in most cases, increase the response to phosphorus and leave the response to carbon (background) fairly unimpaired, thus effectively increasing the selectivity.

It may be mentioned that the concept of choosing a derivative with a particular detector in mind is quite frequently employed in residue analysis. And with the development of more diversified selective detectors, we are sure to see more of it. Thiophosphoryl derivatives of phenols for the flame photometric detector (*59*), nitrophenyl derivatives of amines and thiols (*60*) and brominated anilines for the EC detector (*61*), chloroacetylated phenols for the microcoulometric detector (*62*), and many other examples (*63*) would be worth mentioning. The selectivity of a chemical reaction combined with the selectivity of a gas chromatographic detector can provide superior analytical efficiency.

We shall return to the alkali flame detector in the last part of this presentation, in which I would like to describe some of our recent research efforts. Before we do this, however, I would like to make a few remarks about a very efficient and widely used system, the so-called flame photometric detector.

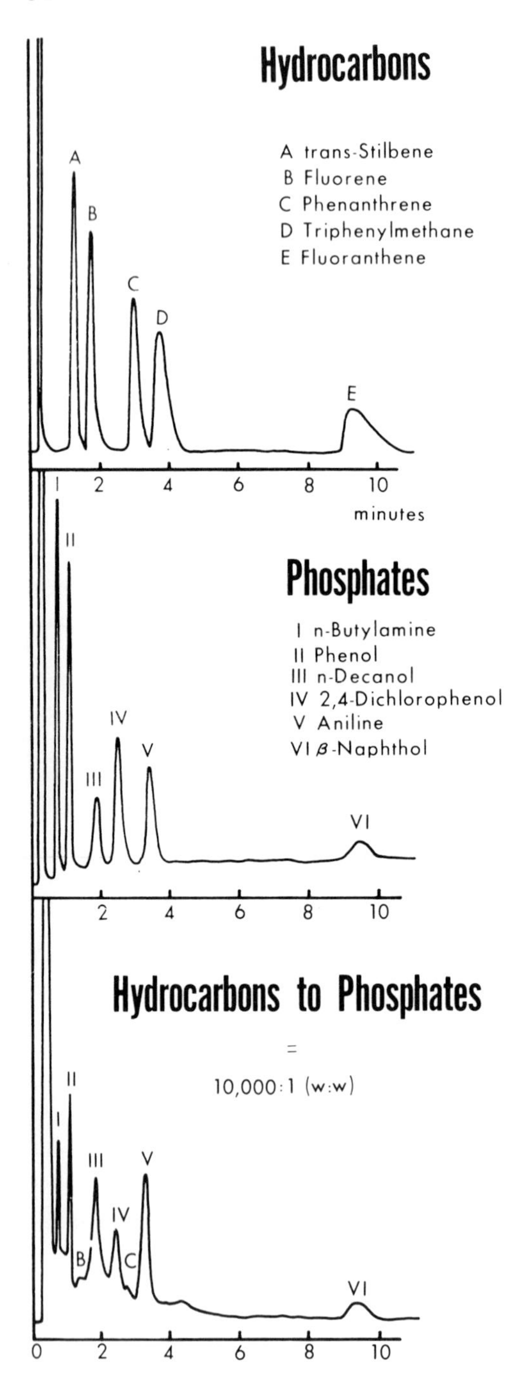

Figure 10. Selectivity of diethylphosphate derivatives vs. *hydrocarbons. Pressed and drilled KCl pellet in a Barber-Coleman FID.*

The Flame Photometric Detector

The FPD is based on a German patent describing the emission obtained with phosphorus and sulfur compounds in a hydrogen-rich flame (*64*). Brody and Chaney developed this analytical method into a detector for gas chromatographic effluents (*65*) and predicted (correctly) its development in the years to come. Today, Tracor, Inc., markets it as "Melpar flame photometric detector" in single- and double-channel versions.

Unlike the AFD, of which almost every residue chemist sports his own modification, FPD's are essentially alike. They are used in increasing numbers and with considerable success.

Their *modus operandi* is simple. Two emission bands can be monitored above the shielded flame by a combination of a narrow band-pass interference filter and a suitable photomultiplier tube. These emission bands at 526 and 394 nm are believed to originate from HPO and SS species. While phosphorus compounds cause only negligible response in the "sulfur mode" (*i.e.*, at 394 nm), sulfur compounds can simulate lower amounts of phosphorus in the "phosphorus mode" (526 nm). Sulfur can, in fact, produce up to one fourth the response of an equal weight of phosphorus, depending on concentration; however, the correct assignment of P or S content is no problem when the response in both modes is compared (*66*).

The flame photometric detector possesses a few characteristics which predestine it for residue analysis. As shown mainly by the group of Bowman, Beroza, and coworkers (*67*, with literature references), it discriminates against compounds devoid of phosphorus or sulfur by factors in the range of four to five orders of magnitude and can therefore be used for samples with little history of purification.

Another advantageous characteristic is the FPD's compatibility with temperature programming. The photomultiplier part of the detector is sensitive to heat, though; from column temperatures between 180° and 200° up, a serious noise problem can arise. However, a few copper cooling coils soldered around the metal part of the detector can improve the situation considerably (*68*).

The flame has to be reignited after each solvent peak, unless the solvent is vented or an automatic relighting system used. Although a nuisance, this does not detract significantly from the detector's value.

The sensitivity of the FPD is considerably lower than that of the AFD; however, there is usually ample sample available in residue analysis and selectivity becomes the more important quality. Minimum detectable limits vary in the reports, perhaps because of the use of different compounds. Thus, two papers from Tracor report 200 and 40 pg (*66*) and

40 and 10 pg (*69*) of elemental sulfur and phosphorus, respectively. Brody and Chaney's original report (*65*) mentioned 250 pg of parathion. Addison and Ackman, who chromatographed elemental phosphorus, reported the significantly lower figure of 1 pg (*70*).

Generally, the residue chemist should be able to detect 1 ng of parathion in the phosphorus mode easily; the linear range for P should be around three orders of magnitude.

The FPD with dual photomultiplier arrangement and dual channel output can monitor P and S simultaneously; if the appropriate circuitry is added, it can also record flame conductivity. Such an arrangement invites studies of P/S ratios; it can also answer the question whether a particular peak in the chromatogram contains P, or S, or both.

Bowman and Beroza, who described a dual channel system, reported that the response ratio—$R(P)/\sqrt{R(S)}$ because of the exponential calibration curve for sulfur—could be used to distinguish PS, PS_2, and PS_3 compounds. "Thus, the response ratio of a given compound is approximately equal to the response ratio of a PS compound multiplied by the atomic ratio of phosphorus to sulfur in a given molecule" (*67*).

Grice *et al.*, with a three-channel FPD/FID, used the signal/noise values in FID/FPD ratios to distinguish hydrocarbons (characterized by very large numbers) from P or S compounds (fractional numbers) (*66*). Quite recently, an interesting paper by Mizany reported that a O_2/H_2 ratio of 0.4 to 0.5—which is quite different from that recommended by the manufacturer—produces higher response for sulfur compounds, whose chemical structure (oxidation state) furthermore significantly influences the signal magnitude (*71*).

Some Comparisons

In many aspects, the alkali flame and the flame photometric detectors compete with each other for the favor of the residue chemist. This competition is reflected by the two major firms which produce these detectors and by the two big U.S. agencies who use them.

The FID, of course, is a detector *sui generis*—but the AFD and the FPD can well be compared on several counts. The sensitivity of the AFD for phosphorus compounds can be two orders higher. This demands, however, a clean alkali-surface and optimized conditions which are difficult to achieve with some modifications. The work by Berck *et al.* (*72*) on the determination of phosphine by three detectors demonstrates this rather clearly.

Top selectivity, on the other hand, is claimed by the FPD (*67, 69*)—although the AFD, when at optimum performance, runs not far behind.

In my opinion, the FPD is easier to adjust and keep at optimum performance. This may be decisive in such operations as air quality control (*73*). However, it is far more expensive than the AFD, which can be made from an FID within less than an hour.

The FPD performs generally better with temperature programming but produces inordinate noise at high temperature unless suitably cooled. The flame has to be reignited after each injection. The AFD, on the other hand, is prone to fluctuations in baseline and sensitivity, especially when the alkali surface becomes contaminated. The FPD can be run with settings recommended by the factory, as do the commercial AFD's. In both cases, however, different flow rates are favored by some authors (*71, 74*). Lab-made AFD's, on the other hand, definitely need optimization in flow rates (and electrode position if possible), but can give superior results without too much effort. So far, the general picture with the special emphasis on phosphorus.

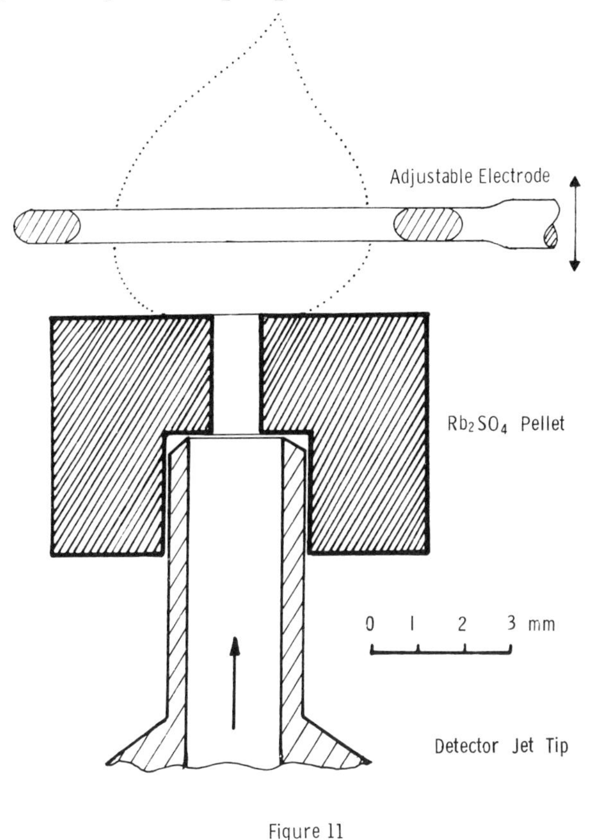

Figure 11. Alkali-flame detector modification (Ref. 75*)*

This picture changes with sulfur. Although the AFD can be used for sulfur determinations (55), the FPD is far superior in selectivity and clearly the detector of choice.

The FPD does not respond to nitrogen or halogens; the AFD does, but much less so than to phosphorus. In those areas, the competition comes primarily from other types of detectors. These are the micro-coulometric and the electrolytic conductivity detectors in the nitrogen and halogen areas and, of course, the electron capture detector for chlorinated hydrocarbons and other electron-attaching compounds.

We found the AFD, for instance, suitable for the analysis of *s*-triazine herbicides, but were plagued by temperamental detector performance (52). Hartmann recently described an AFD of similar construction with vastly improved performance (35). In my opinion, the electrolytic conductivity detector has at present the best chances of "making it" in the nitrogen area, but the matter is far from settled.

The AFD is no match for the electron capture detector in the analysis of highly halogenated compounds when high sensitivity and good quantitation are demanded. Neither are the two other detectors. On the other

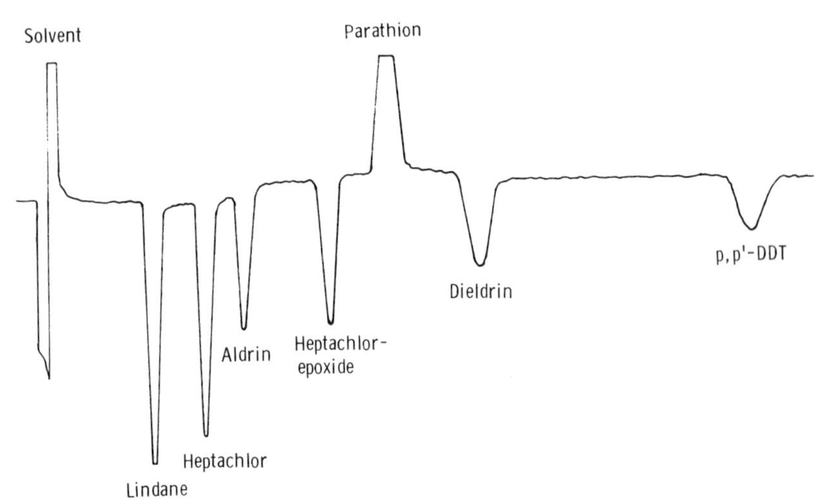

Figure 12. Electrode: 7 mm i.d., set 1.5 mm above the pellet. Pellet: pressed Rb_2SO_4, 1-mm bore. Potential: +240 Volts. Flow, ml/min.: H_2 33, N_2 50, Air 215 (Ref. 75).

Column: 9.8% DC-200 + 15.8% QF-1 (50/50 w/w) on Anakrom ABS, 90/100 mesh, prepared in fluidized bed. Column bath 190°, injection port 220°, detector 210°

Injection: one microliter of a hexane solution containing 10 ppm of each of the chlorinated hydrocarbons and 1 ppm of parathion.

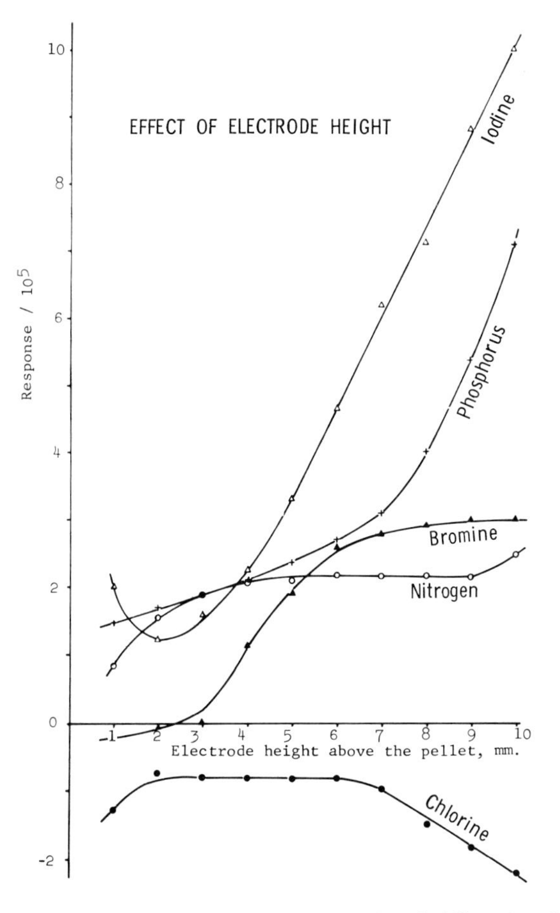

Journal of Chromatography

Figure 13. Electrode: 7-mm i.d. Pellet: pressed Rb_2SO_4, 1-mm bore. Potential: +240 Volts. Flow, ml/min.: H_2 33, N_2 50, Air 215 (Ref. 75).

Injections: one microliter of hexane solutions containing 1 μl/ml each of chlorobenzene, bromobenzene, iodobenzene, and benzylamine and 10 μg/ml of tri-n-butyl phosphate.

hand, the ECD is quite sensitive to background materials of all sorts, and for the "fast and dirty" analysis, other detectors may be more suitable.

The AFD is selective to arsenic (*53*)—also the microwave plasma emission detector (*54*)—but the response is poor. For practical As determination, the colorimetric or neutron activation techniques are preferable.

This short comparison concludes the review of the three flame detectors. Now I would like to describe some recent observations made by our group on the alkali flame detector in general and its negative response (inverted peaks) in particular.

The Alkali Flame Detector Revisited

In discussing the dual-channel flame photometric detector, I praised its ability to determine the phosphorus/sulfur ratio in chromatographic peaks. "Peaks" was meant quite literally, because a peak, most probably one among many, is all the evidence of a particular compound one may have. More and more often, the analyst is confronted with this very type of problem. The sample may be a soil or plant containing pesticide decomposition products, it may be a biomedical extract with drug metabolites, it may be a concentrate of organic substances polluting air or water, it may be an extract from lunar dust causing peaks in the FID—in fact, it may be a dozen types of things, all complex and often available only in small quantities. It is these problems which demand detectors capable of giving some structural information on unknowns, which are separated from other unknowns (for all practical purposes) on the strip

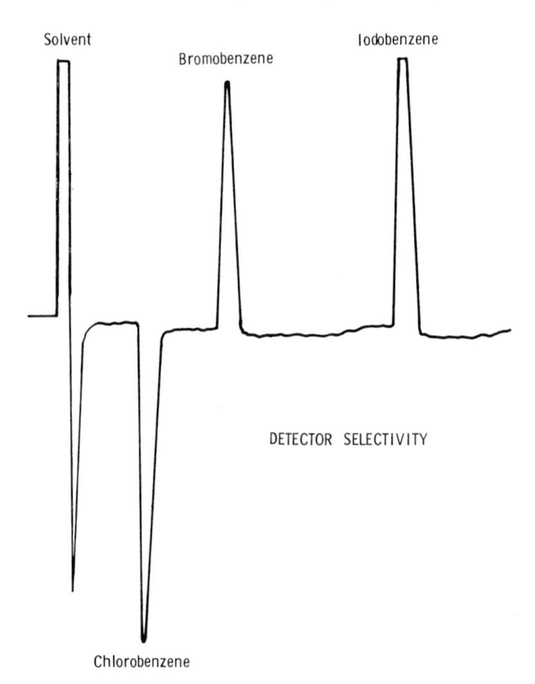

Figure 14. Electrode: 7-mm i.d. set 4 mm above the pellet, +240 Volts. Pellet: pressed Rb_2SO_4, 1-mm bore. Flow, ml/min.: H_2 33, N_2 50, Air 215 (Ref. 75).

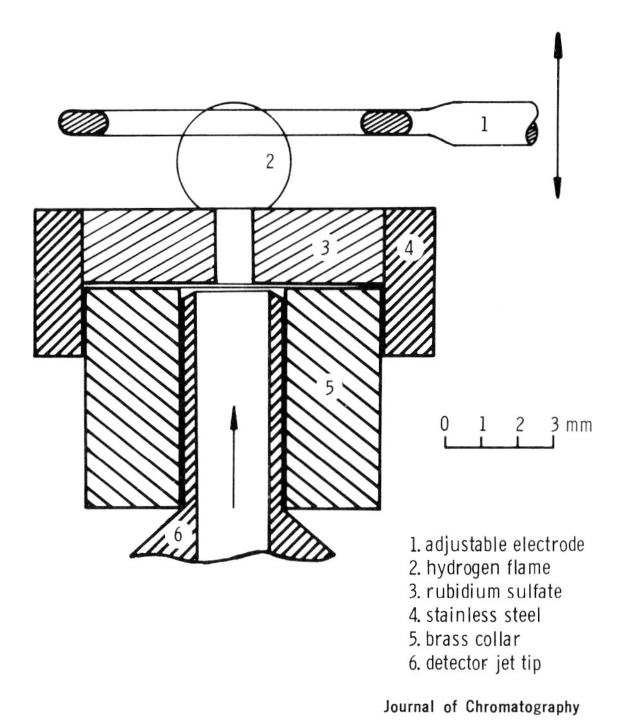

Figure 15. Modified alkali-flame detector (Ref. 76)

chart recorder only. The alkali-flame is a good example of such a detector.

The AFD, in addition to its already temperamental behavior, will often show negative response (inverted peaks). Such peaks have been observed by many authors, who probably considered them an interesting nuisance (*24, 25, 31, 32, 33, 43*). There appeared only two reports in the literature which suggested or made some use of negative response: Dressler and Janák's paper on sulfur compounds (*55*) and our own on chlorinated hydrocarbons (*75*).

For these studies, we used a pressed and drilled pellet made from rubidium sulfate of 99.9% purity, as shown in Figure 11. The collector electrode had an inner diameter of 7 mm and could be moved vertically between 1 and 10 mm above the pellet. Figure 12 shows the analysis of a chlorinated hydrocarbon mixture, after the detector had been optimized for negative response. Included in the chromatogram is parathion, which gives a positive peak. Using the particular flow rates best suited for negative chlorine response, we checked some of the other alkali flame-active elements by varying the electrode height as shown in Figure 13. Obviously, the AFD has the capability to distinguish between elements on ground of their response profiles, or, at the right conditions, through the direction of their peaks. This is shown in Figure 14 with the detector set to distinguish chlorine from bromine and iodine. Incidentally, this is the more difficult chromatogram to obtain; it is much easier to keep chlorine and bromine negative and let only iodine be positive.

These results (*75*) brought about a subsequent study (*76*) in which we attempted to define the ranges of negative response for Cl, Br, I, N, and C. Phosphorus was also run for purposes of comparison; however, its response remained positive under the particular conditions used.

The alkali flame detector had a slightly different configuration, designed to allow frequent bead changes without scratching or contaminating the alkali surface. A smooth and clean pellet is necessary in order to obtain adequate negative response. The pellet is held in a stainless steel ring (Figure 15), the same ring in which it was originally pressed. It is kept in the proper position by a collar, rather than resting on the jet tip. While this modification does not seem to alter response characteristics to any significant extent, it represents a more stable configuration.

Four different parameters were varied: the electrode height above the alkali surface, the alkali pellet bore, the hydrogen flow, and the nature and the flow of the carrier gas. While the detailed results are too involved to be presented in this short review, I would like to illustrate some of the changes in response direction and magnitude.

Figure 16 shows three halides (chloro-, bromo-, and iodobenzene), determined with pellets of different bead diameter, in their dependence

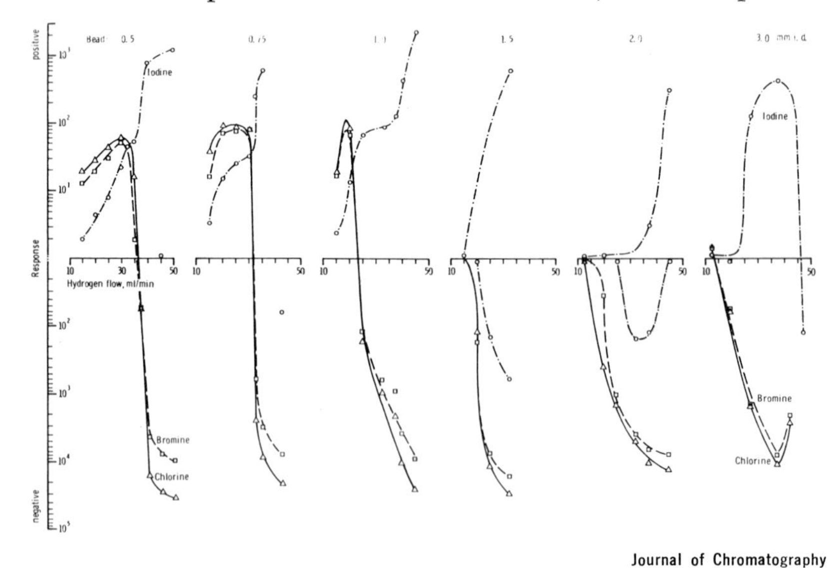

Figure 16. Halogen response profiles at various hydrogen flows and bead bores (Ref. 76)

Electrode height, 2 mm. One-μl triplicate injections of a 1% solution of chlorobenzene, bromobenzene, and iodobenzene. N_2 50 ml/min.

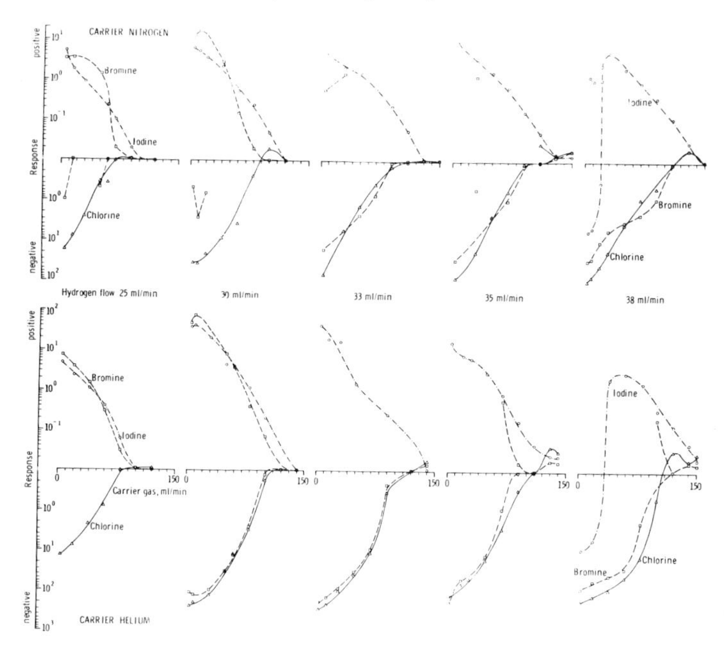

Figure 17. Halogen response profiles for varying carrier gas flow (nitrogen and helium) and five selected hydrogen flows (Ref. 76)

Electrode height 2 mm, bead bore 1.0 mm. One-μl duplicate injections of 0.01% chlorobenzene, bromobenzene, and iodobenzene.

on the hydrogen flow. Split peaks are observed in some cases (especially in the transition areas between negative and positive response) and are recorded as two distinct data, one for the positive, one for the negative part of the peak. As bead bore and hydrogen flow increase, so does the area of negative response.

Figure 17 shows the halogen response for a constant bead bore and different hydrogen flows in its dependence on the carrier gas flow; both helium and nitrogen were used. We chose these two gases because it had been reported that helium increases the AFD's response (*44, 53, 77*). It is interesting to pursue the turnabout of the halogens with increasing hydrogen flow. In the beginning, only chlorine is negative. Then, with a few relapses shown as split peaks, bromine joins the negative side. Finally, iodine begins to switch over.

Figure 18 shows a similar plot with one nitrogen compound (aniline) and a series of carbon–hydrogen–oxygen compounds of various structures. Only *n*-decane has been represented by a curve, since the other compounds fall closely around it. The aniline's six carbon atoms make it join the carbon compounds in areas of negligible nitrogen response.

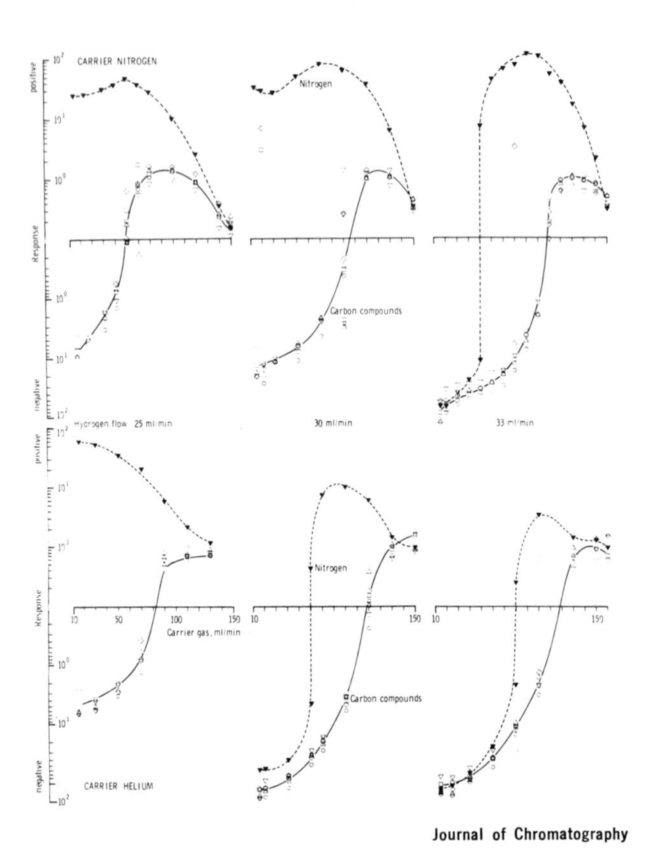

Journal of Chromatography

Figure 18. Nitrogen and carbon response profiles for varying carrier gas flow (nitrogen and helium) and three selected hydrogen flows (Ref. 76)

Electrode height 2 mm, bead bore 1.0 mm. One-μl single injections of 1% aniline and 1% each of p-*xylene,* n-*decane,* p-*cymene 1-octanol, and anisole.*

Interestingly enough, nitrogen as a carrier gas performs better for nitrogen compounds than helium does.

Figure 19 shows the positive phosphorus response, again with two different carrier gases. No significant difference between the carrier gases is noticed in our particular AFD version. The general shapes of the curves are similar, although their position shifts, of course, toward higher response with an increase in hydrogen flow, when data obtained at the same carrier gas flow are compared.

Figure 20 displays some chromatograms taken from these serial measurements. The flow rates are listed in Table I. If one particular analysis is desired, the electrode height and other parameters (air, for instance) are easily optimized, resulting in more esthetically pleasing chromatograms.

It may not be amiss at this point to recommend a certain measure of caution in measuring, reporting, or using selectivity parameters of one definition or another. Although such data are extremely useful in planning an analytical approach or describing a particular detector's characteristics, they can also be misleading. This is especially true in the case of a detector as multi-faceted as the AFD. Figures 16 through 20 may be taken as illustrations to the difficulties encountered in evolving an adequate description of "selectivity." In this case, the problem is, furthermore, aggravated by the occurrence of negative response.

In general, negative response is favored by, or only obtainable through, a smooth, clean, and homogeneous alkali surface, a large bead bore, a high hydrogen flow, and a low carrier gas supply. The last two conditions also produce a hotter flame in contact with a larger alkali surface area, consequently a larger response in general, be it positive or negative.

Again, as a general statement—which may well not apply to other detector modifications or conditions—the ease of obtaining negative response decreases in the order Cl–Br–I, and N–P. It would be interesting to compare C–Si in this respect, in a configuration where SiO_2 deposits do not contaminate the alkali surface.

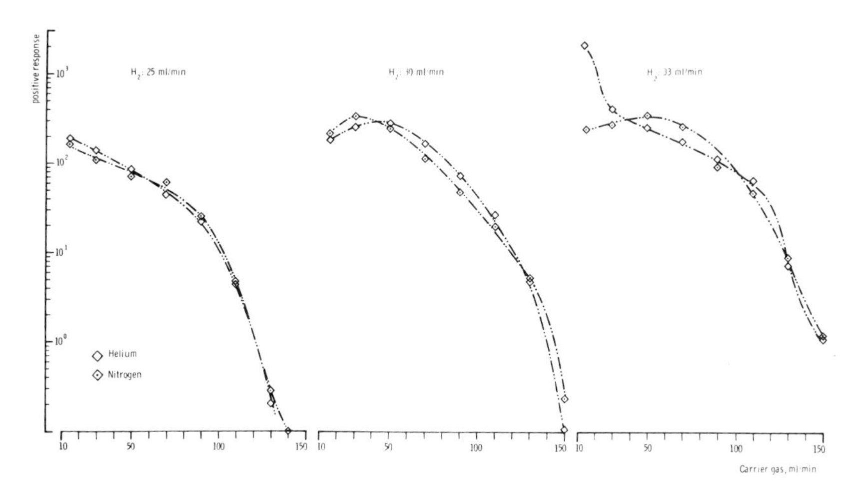

Journal of Chromatography

Figure 19. Phosphorus response profiles for varying carrier gas flow (nitrogen and helium) and three selected hydrogen flows (Ref. 76)

Electrode height 2 mm, bead bore 1 mm. One-μl triplicate injections of 0.001% trimethylphosphate.

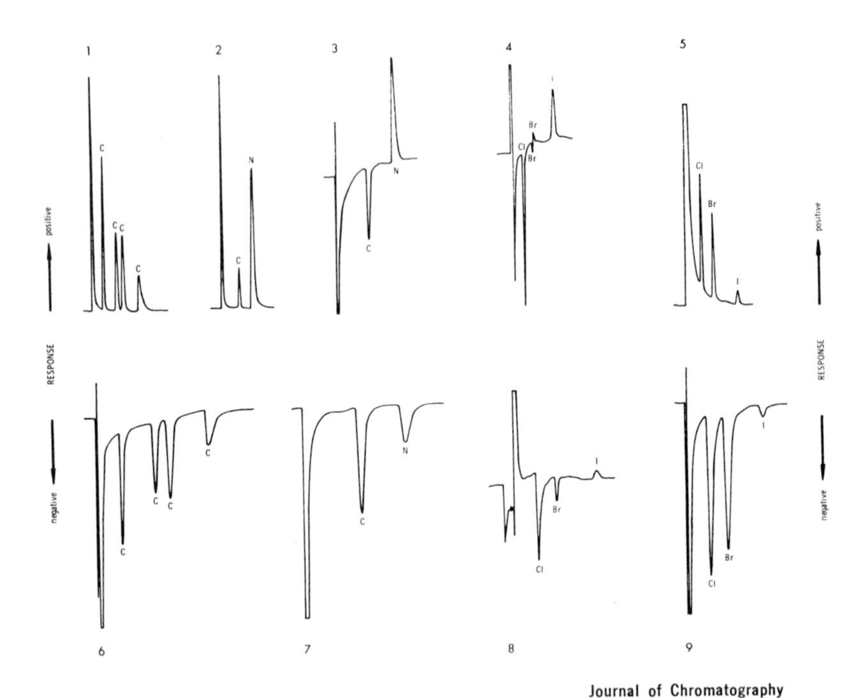

Figure 20. Typical chromatograms at various flow rates (Ref. 76)

Bead bore 1.0 mm; electrode heights 6 mm (#5) and 2 mm (all others)

Interesting as the qualitative analysis of heteroatoms by the AFD may be, it is not necessarily applicable to complex mixtures. To distinguish heteroatoms by their direction of response may necessitate detector settings which do not represent the best sensitivity obtainable, and are consequently more susceptible to background interference. Furthermore, since the carrier gas flow may be dictated by the detector, the column dimensions have to be adjusted accordingly.

Another potentially valuable but largely unexploited characteristic of the alkali flame detector is its linear response to the amount of heteroatom introduced into the flame. There has been some controversy in the literature on this point. Cremer (58), Karmen (57), Giuffrida (53), and others hold that the response is linear for the elements they investigated.

Table I. Flow Rates in Figure 20

	Ml/Min.								
	1	*2*	*3*	*4*	*5*	*6*	*7*	*8*	*9*
N_2(He)	92	92	46	78	50	37	(27)	42	(43)
H_2	26	26	33	30	16	33	30	37	37

Janák obtained a different relationship for halides (*27*) and Hartmann expressed some doubts on nitrogen compounds (*35*).

We have tried to investigate (*78*) this relationship with a variety of compounds containing chlorine, bromine, iodine, nitrogen, or phosphorus, some in negative, some in positive mode. We always obtained a roughly linear correlation; data points which fell outside the line did not reveal any particular trend. Figure 21 shows an example of such a series, where equal weights of chlorine-containing compounds were chromatographed. Not under all conditions, however, did the extrapolated lines go through the origin. A change in conditions, for instance in the electrode height, usually improved the situation. The reason for such a line not going through zero could lie in a concentration effect or, more likely in some cases, in the influence of carbon present in the molecules. We have not followed up on this problem but simply circumvented it by choosing different conditions.

Why did we want the line to go through the origin? Mainly because this would allow us to estimate the % heteroatom content of an unknown

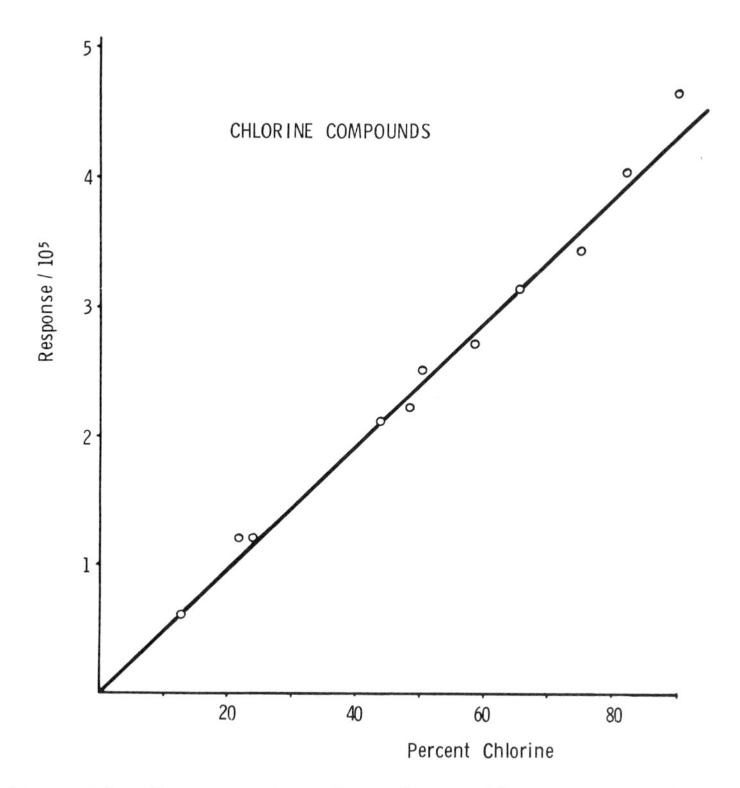

Figure 21. Response of equal weights of chlorine compounds. Conditions as given in Figure 8 except electrode height 1.5 mm.

Table II. Determination

Compound	*% X, Theor.*	*% X, Calc.*
1, 5–Dichloropentane	50.3	
a, a–Dichlorotoluene	44.1	44.0
p–Dichlorobenzene	48.3	
1, 2, 3–Trichloropropane	72.2	73.5
1, 5–Dichloropentane	50.3	
s–Tetrachloroethane	84.6	84.2
1, 2, 3–Trichloropropane	72.2	
1, 2, 4–Trichlorobenzene	58.7	58.7
Benzyl chloride	28.0	
Hexachloro–1, 3–butadiene	81.7	82.7
1, 2, 3–Trichloropropane	72.2	
1, 5–Dichloropentane	50.3	55.4
Hexachloroethane	90.0	
p–Dichlorotoluene	41.1	43.7
3–(Chloromethyl)heptane	23.9	
Hexachloroethane	90.0	86.5
1, 2, 4, 5–Tetrachlorobenzene	65.8	
1–Chloronaphthalene	21.8	28.0
a, a–Dichlorotoluene	44.1	
Benzyl chloride	28.0	26.4
1, 5–Dichloropentane	50.3	
1, 4–Dichlorobutane	55.8	60.0
a, a, a–Trichlorotoluene	54.5	
1–Chloronaphthalene	21.8	27.4

compound with just one more standard compound to use—provided we knew the injected weight of the unknown.

While such an approach would prove adequate for a pure compound —barring decomposition in the column and similar adversities—unknowns contained in a mixture cannot be handled this way. What is needed is one other bit of information concerning the approximate amount of compound represented by a particular peak. This amount can be obtained —very roughly—by using a second, nonselective, detector such as the FID, preferably in a different instrument in order to avoid contamination by alkali salts.

Suppose we examine the same mixture containing two chlorine compounds *a* and *b* with different % chlorine content by both the AFD and the FID. Assuming that the FID responses R_a and R_b represent the respective carbon fractions (which becomes less correct with increasing

of Halide Content[a]

Compound	*% X, Theor.*	*% X, Calc.*
Ethylene dibromide	85.3	
1–Bromohexane	48.4	44.8
Bromobenzene	51.0	
Bromocyclohexane	49.0	48.8
1, 2–Dibromopropane	79.5	
1–Bromopentane	53.0	51.4
Dibromomethane	91.9	
1–Bromopentane	53.0	51.9
Ethylenedibromide	85.3	
2–Bromopentane	53.0	50.8
Bromoform	94.8	
Bromocyclohexane	49.0	46.3
Bromobenzene	51.0	
1, 3–Dibromopropane	79.5	81.6
Ethylenedibromide	85.3	
1–Bromo–3–methylbutane	53.0	49.0
1–Bromo–2–methylpropane	48.7	
1–Bromo–3–methylbutane	53.0	49.1
2–Iodopropane	74.7	
1–Iodopropane	74.7	72.5
p–Iodotoluene	58.2	
Iodobenzene	62.2	61.3

[a] Two-microliter single injections of 0.1 or 0.01% (w/v) solutions containing two halides.

1. *Alkali flame detection:* Rb_2SO_4 pellet, 1-mm bore; collector electrode 7-mm i.d., 2 mm above the pellet, at +240V. Flow rates in ml/min: H_2 38, N_2 13, Air 250. Modified Barber-Colman 5320.

2. *Flame ionization detection:* Flow rates in ml/min: H_2 45, N_2 33, Air 300. Mikrotek MT 220. *Columns:* 10% OV–17 on Chromosorb W, HP, 80/100 mesh, 4-mm i.d. × 150 (180) cm borosilicate glass.

oxygen content) and the AFD the respective chlorine fractions, one can estimate the Cl/C ratio from a formula like

$$\left[\frac{\%\mathrm{Cl}}{\%\mathrm{C}}\right]_a = \left[\frac{R_a}{R_b}\right]_{AFD} \left[\frac{R_b}{R_a}\right]_{FID} \left[\frac{\%\mathrm{Cl}}{\%\mathrm{C}}\right]_b$$

Assuming that the compounds in question were similar in structure (as, for instance, chlorinated hydrocarbons), one could further take the carbon response as representative of the rest of the molecule and substitute %Cl/%C by %Cl/(100 − %Cl). If the standard is chosen relatively close to the unknown in structure and chlorine content, a rough estimate results. Table II shows a few calculations done in this manner for pairs of halogen compounds.

The intrinsic difficulties in such an approach become forbidding, if not insuperable, when polar compounds, or compounds containing more than one type of alkali flame-active element, are to be analyzed against complex backgrounds. Hopefully, the use of other selective detectors—*e.g.*, some of those discussed in this symposium—will show a solution to the problem.

It is my belief that a lot of future research will be directed along lines similar to those we have attempted to follow. This research has merely touched the periphery of a new field called qualitative and quantitative structure analysis by chromatography.

There exist strong demands for sensitive detectors with capabilities for structure analysis; demands from people or agencies concerned with food quality, environmental quality, successful (and safe) agriculture, human health, extraterrestrial materials, forensic and archeological artifacts, and so forth *ad infinitum*. Future flame detectors will certainly be formed by these demands. There should be exciting years of research lying ahead.

Acknowledgment

I gratefully acknowledge my collaborators whose research I have mentioned in this manuscript. They are Klaus Gerhardt, Stanislaw Lakota, Gerhard Ertingshausen, David L. Stalling, Roy Koirtyohann, Robert F. Moseman, Roger C. Tindle, Lyle D. Johnson, and Charles D. Ruyle. My special thanks go to Charles W. Gehrke for his encouragement and cooperation and to Edward E. Pickett for his assistance on spectroscopic problems. This paper is a contribution from the Missouri Agricultural Experiment Station, Journal Series No. 6039, approved by the director. Research by the author's group has been funded by USPHS grant 5RO1-FD-00262-03, formerly CC-00314-02 and by USDA-ARS grant 12-14-100-9146. This study has been assisted by the Environmental Trace Substance Center, Columbia, Mo.

Literature Cited

(1) Thornberg, W., Beckman, H., *Anal. Chem. (Annual Review)* (1969) **41,** 140 R.
(2) Dean, J. A., Morrow, R. W., Shultz, W. D., *Mid-America Symposium on Spectroscopy, Chicago, Ill., May 1969.*
(3) Gunther, F. A., Lopez-Roman, A., Asai, R. I., Westlake, W. E., *Bull. Environ. Contam. Toxicol.* (1969) **4,** 202.
(4) Bowman, M. C., Beroza, M., *J. Chromatog. Sci.* (1969) **7,** 484.
(5) Gutsche, B., Herrmann, R., *Z. Anal. Chem.* (1969) **245,** 274.
(6) Nowak, A. V., Malmstadt, H. V., *Anal. Chem.* (1968) **40,** 1108.
(7) Gilbert, P. T., *Anal. Chem.* (1966) **38,** 1921.
(8) Gutsche, B., Herrmann, R., *Z. Anal. Chem.* (1970) **249,** 168.

(9) Umstead, M. E., Woods, F. J., Johnson, J. E., *ACS–CIC Joint Conference, Toronto, Canada, May 1970.*
(10) Scolnick, M., *Intern. Symp. Advan. Chromatog., 6th, Miami Beach, Fla., June 1970.*
(11) Cremer, E., Kraus, T., Bechtold, E., *Chem. Ing. Tech.* (1961) **33,** 632.
(12) Guichard, N., Buzon, J., "Manuel Pratique de Chromatographie en Phase Gazeuse," pp. 155–196, J. Tranchant, Ed., Masson et Cie., Paris, 1968.
(13) Johns, T., Sternberg, J. C., "Instrumentation in Gas Chromatography," pp. 179–207, J. Krugers, Ed., Centrex, Eindhoven, 1968.
(14) Fohner, O. F., Haase, D. J., *Anal. Chim. Acta* (1969) **48,** 63.
(15) Gill, J. M., Hartmann, C. H., *J. Gas Chromatog.* (1967) **5,** 605.
(16) Knapp, A., *Chromatographia* (1969) **2,** 67 and 111.
(17) Nunnikhoven, R., *Z. Anal. Chem.* (1968) **236,** 79.
(18) Feldman, G., *cited in Gas-Chrom. Newsletter* (March/April 1970) **11** (2), 4, Applied Science Labs, State College, Pa. 16801.
(19) Moseman, R. F., Aue, W. A., *J. Chromatog.* (1970) **49,** 432.
(20) Brazhnikov, V. V., Gurév, M. V., Sakodynsky, K. I., *Chromatog. Rev.* (1970) **12,** 1.
(21) Rice, C. W., U. S. Patent **2,550,498** (1944).
(22) Giuffrida, L., Ives, F., *J. Assoc. Offic. Agr. Chemists* (1964) **47,** 1112.
(23) Moesta, H., Schuff, P., *Ber. dt. Bunsenges.* (1965) **69,** 895.
(24) Karmen, A., *Intern. Symp. Advan. Chromatog., 5th, Las Vegas, Nev., January 1969.*
(25) Karmen, A., *J. Chromatog. Sci.* (1969) **7,** 541.
(26) Nowak, A. V., *Dissert. Abstr.* (1968) **29,** No. 2, 510-B.
(27) Janák, J., Svojanovsky, V., Dressler, M., *Coll. Czech. Chem. Commun.* (1968) **33,** 740.
(28) Saturno, J. J., Cooke, W. D., "Abstracts of Papers," *153rd ACS Meeting, New York* (1966).
(29) Page, F. M., Woolley, D. E., *Anal. Chem.* (1968) **40,** 210.
(30) Brazhnikov, V. V., Gurév, M. V., Sakodynsky, K. I., *Chromatographia* (1970) **3,** 53.
(31) Abel, K., Lanneáu, K., Stevens, R. K., *J. Assoc. Offic. Agr. Chemists* (1966) **49,** 1022.
(32) Hartmann, C. H., *Bull. Environ. Contam. Toxicol.* (1966) **1,** 159.
(33) Hartmann, C. H., *Aerograph Research Notes* (Spring, 1966).
(34) Dimick, K. P., Hartmann, C. H., Oaks, D. M., Trone, E., U. S. Patent **3,423,181** (1969).
(35) Hartmann, C. H., *J. Chromatog. Sci.* (1969) **7,** 163.
(36) Jentzsch, D., Zimmermann, H. G., Wehling, I., *Z. Anal. Chem.* (1966) **221,** 377.
(37) "Für die Praxis," 35 GC, Bodenseewerk Perkin Elmer & Co., Überlingen, West Germany, 1967.
(38) Strictly confidential information.
(39) Giuffrida, L., *J. Assoc. Offic. Agr. Chemists* (1964) **47,** 293.
(40) Giuffrida, L., U. S. Patent **3,372,994** (1968).
(41) Wessel, J. R., *J. Assoc. Offic. Agr. Chemists* (1968) **51,** 666.
(42) Coahran, D. R., *Western ACS Conference, Corvallis, Ore., June 1965.*
(43) Dressler, M., Janák, J., *Coll. Czech. Chem. Commun.* (1968) **33,** 3970.
(44) Ebing, W., *Chromatographia* (1968) **1,** 382.
(45) Aue, W. A., Ertingshausen, G., "Abstracts of Papers," *154th ACS Meeting, Chicago, Ill.* (1967).
(46) Aue, W. A., Gehrke, C. W., "Abstracts of Papers," *152nd ACS Meeting, New York* (1966).
(47) Mees, R. A., Spaans, J., *Z. Anal. Chem.* (1969) **247,** 252.
(48) Tindle, R. C., Aue, W. A., Gehrke, C. W., *AOAC Meeting, 80th, Washington, D. C., October 1966.*

(49) Bostwick, D. C., Giuffrida, L., *J. Assoc. Offic. Agr. Chemists* (1967) **50,** 577.
(50) Ertingshausen, G., Gehrke, C. W., Aue, W. A., *Separ. Sci.* (1967) **2,** 681.
(51) DeLoach, H. K., Hemphill, D. D., *J. Assoc. Offic. Agr. Chemists* (1969) **52,** 553.
(52) Tindle, R. C., Gehrke, C. W., Aue, W. A., *J. Assoc. Offic. Agr. Chemists* (1968) **51,** 682.
(53) Ives, N. F., Giuffrida, L., *J. Assoc. Offic. Agr. Chemists* (1967) **50,** 1.
(54) Johnson, L. D., Aue, W. A., *Midwest Regional ACS Meeting, 5th, Kansas City, Mo., October 1969.*
(55) Dressler, M., Janák, J., *J. Chromatog. Sci.* (1969) **7,** 451.
(56) Bechtold, E., PhD thesis, University of Innsbruck, Austria (1962).
(57) Karmen, A., *Anal. Chem.* (1964) **36,** 1416.
(58) Cremer, E., Moesta, H., Hablik, K., *Chem. Ing. Tech.* (1966) **38,** 580.
(59) Bowman, M. C., Beroza, M., *J. Assoc. Offic. Agr. Chemists* (1967) **50,** 926.
(60) Crosby, D. G., Bowers, J. B., "Abstracts of Papers," *153rd ACS Meeting, Miami Beach, Fla.* (1967).
(61) Gutenmann, W. H., Lisk, D. J., *J. Agr. Food Chem.* (1964) **12,** 46.
(62) Chin, W-T., Cullen, T. E., Stanovick, R. P., *J. Gas Chromatog.* (1968) **6,** 248.
(63) Symp. Chem. Deriv. Pesticide Residue Anal., "Abstracts of Papers," *155th ACS Meeting, San Francisco, Calif.* (1968).
(64) Draeger, H., Draeger, B., W. German Patent **1,133,918** (1962).
(65) Brody, S. S., Chaney, J. E., *J. Gas Chromatog.* (1966) **4,** 42.
(66) Grice, H. W., Yates, M. L., David, D. J., *J. Chromatog. Sci.* (1970) **8,** 90.
(67) Bowman, M. C., Beroza, M., *Anal. Chem.* (1968) **40,** 1448.
(68) Stalling, D. L., Fish Pesticide Residue Laboratory, Columbia, Mo., private communication.
(69) O'Donnell, J. F., *Am. Lab.* (February 1969).
(70) Addison, R. F., Ackman, R. G., *J. Chromatog.* (1970) **47,** 328.
(71) Mizany, A. I., *J. Chromatog. Sci.* (1970) **8,** 151.
(72) Berck, B., Westlake, W. E., Gunther, F. A., *J. Agr. Food Chem.* (1970) **18,** 143.
(73) Stevens, R. K., O'Keeffe, A. E., *Anal. Chem.* (1970) **42,** 143A.
(74) Sissons, D. J., Telling, G. M., *J. Chromatog.* (1970) **47,** 328.
(75) Lakota, S., Aue, W. A., *J. Chromatog.* (1969) **44,** 472.
(76) Gerhardt, K. O., Aue, W. A., *J. Chromatog.* (1970) **52,** 47.
(77) Ford, J. H., Beroza, M., *J. Assoc. Offic. Agr. Chemists* (1967) **50,** 601.
(78) Aue, W. A., Gerhardt, K. O., Lakota, S., unpublished material.

RECEIVED June 27, 1970.

5

Gas Chromatographic Measurement and Identification of Pesticide Residues with Electron Capture, Microcoulometric, and Electrical Conductivity Detectors

WILLIAM E. WESTLAKE

Department of Entomology, University of California, Riverside, Calif. 92502

The gas chromatograph is not capable of giving specific qualitative information but can give a high degree of assurance of the identity of the compound being detected. The electron capture detector, while unexcelled for detecting exceedingly minute quantities of many compounds, is virtually useless for qualitative identification. The microcoulometric detection system has a high degree of specificity and is widely used for the confirmation of recorder responses using the electron capture when the presence of an organochlorine pesticide is indicated. The detector is almost specific for chlorine, sulfur, or nitrogen. This specificity, together with retention time data, offers a high degree of assurance of the identity of the compound being measured. The electrolytic conductivity detector is similarly specific and is capable of measuring organic iodine or bromine.

Some remarks of a general nature regarding gas chromatography are appropriate before beginning a discussion of specific detection systems. This paper is not intended to be a review, but references are cited where it seems appropriate to support particular points. The intent is to summarize, in a general way, what the detectors will and will not do and to make known the author's ideas regarding some advantages and disadvantages in their use.

Although some may think otherwise, a substantial majority of analytical chemists will surely agree that, strictly speaking, the gas chromatograph is not an instrument for qualitative analysis. As a quantitative tool it is often superb, and for the isolation of individual components of a mixture it is frequently indispensable. The development, during the past 10 years, of detectors possessing some degree of specificity has made it possible, frequently, to obtain a reasonable assurance of the identity of the detected compound. None of the so-called specific detectors is infallible, for, although they may have a greatly enhanced response for a particular molecular species, they will also respond to a lesser degree to others. For absolute certainty when determining unknown compounds, therefore, some additional verification is essential.

The availability of detectors (particularly electron capture) capable of giving a signal for as little as a few picograms of some compounds has resulted in the universal and almost exclusive use of the gas chromatograph for measuring pesticide residues. Without the gas chromatograph, much of the research on pesticide residues being conducted today would not be possible because the manpower required to process the large samples and to do the chemical work necessary for other methods of analysis for residues in the 0.1 ppm and lower range does not exist. Many of the measurements being reported could not be made at all (at least, not in any practical manner) by means other than gas chromatograph. This latter accomplishment is viewed with mixed emotions by many experienced residue chemists, for the efforts directed toward detecting less and less and toward miniaturizing samples have long since reached the point of no return. The errors inherent in sampling, processing, instrumentation, and operator inconsistencies increase rapidly as levels go below 0.1 ppm and, at the 1 ppb level, it is doubtful that reproducibility can be as good as $\pm 100\%$. When its limitations are recognized and the data properly interpreted, the gas chromatograph is an almost indispensable instrument for the residue chemist engaged in research and for regulatory agencies that must screen thousands of samples each year for the presence of many different pesticides.

As the title of this paper indicates, it is concerned only with the use of the electron capture, microcoulometric, and electrolytic conductivity detectors for quantitation and identification of pesticide residues. Further background information on gas chromatographic detectors may be found in the paper by Westlake and Gunther (*1*).

Electron Capture Detector

The electron capture detector is the most extensively used, by far, of those employed for pesticide residue determinations. It is also the worst possible choice for identifying the detected compounds. Introduced by Lovelock and Lipsky (2) in 1960, the use of this detector has revolutionized pesticide residue determinative procedures. The ability to detect and measure exceedingly minute amounts of many pesticides (particularly the organochlorine compounds) in a wide range of substrates caused its immediate acceptance as an analytical tool for pesticide research. This detector, however, has the unfortunate property of giving responses to a host of compounds other than pesticides and is, therefore, virtually useless for identification. When the identity of the compound being measured is known and operating parameters established, it is unexcelled for quantitative investigations.

The limitations of the electron capture detector were quickly recognized, and the search was continued for more suitable detection systems, resulting in the development of several alternate types, each having its advantages and disadvantages. In the meantime, efforts have continued to make the electron capture detector a qualitative tool and to improve its performance and ease of maintenance. The majority of the electron capture detectors use tritium as an ionization source, although strontium-90 was adopted by some manufacturers. Both have the disadvantage of being volatile at elevated temperatures, and operating temperatures are limited to 225°C or less. Even then, there is a gradual loss of radioactivity, and deposits of sample extractives and column packing materials build up rapidly in the detectors, making them inoperable in a relatively short time. The recent introduction of nickel-63 as an ionization source permits operating temperatures up to 400°C, thus preventing the build-up of column effluents in the detector. Unfortunately, there are disadvantages for some designs, including higher minimum detectable amounts and, more important in the writer's opinion, very limited linear ranges of response. One such detector that was tested had a usable linear range of about 100 to 500 pg of aldrin. In this instance, the minimum detectability was actually lower than necessary but the response was so nonlinear outside this range that it was virtually useless. At least two manufacturers now claim to have nickel-63 detectors that have usable linear ranges and minimum detectability equivalent to or better than the tritium detectors. Another type of electron capture detector that will operate at temperatures up to 400°C uses a "nonradioactive" source in the form of an electrical discharge. This detector gives a response similar to that of other electron capture detectors and is being routinely used by a number of laboratories.

Efforts toward using the electron capture gas chromatograph as a qualitative tool have centered first on the use of two or more different columns that hopefully will give different relative retention times for the various eluting peaks and, with the judicious use of standards, some assurance of the identity of unknowns. This procedure can be time-consuming and, while helpful, still may not yield anything positive. A preferable procedure is to chromatograph a sample, then carry out some chemical reaction (*e.g.*, oxidation, reduction, dehydrochlorination, rearrangement, or addition) to form a derivative that will have a different retention time. A comparison with standards treated in the same manner can give informative data. An illustration of this is the conversion of DDT to DDE by dehydrochlorination. Such indirect methods of getting the electron capture gas chromatograph to give qualitative information may produce useful but not conclusive confirmatory data.

There are two areas in pesticide residue determinations where the electron capture detector is without equal. The first is for the screening of samples by regulatory agencies to eliminate from further consideration all but the small percentage that contain above-tolerance levels of some pesticide(s). Once the retention times and detector responses of the sought compounds have been established, samples can be injected, and those not showing a response indicating the presence of illegal quantities of pesticides may be eliminated from further consideration. Those showing possible above-tolerance levels (normally less than 2% of the total) may then be subjected to independent confirmatory methods to establish definitely the identity of the compound(s). Electron capture responses indicating organochlorine pesticides, for example, are often routinely checked by microcoulometric gas chromatography to establish whether or not the compound(s) contain chlorine. The combination of retention time and chlorine content give reasonable assurance of the identity of the compound. Organophosphorus pesticides may be checked using a phosphorus-sensitive detector, nitrogen-containing pesticides with a nitrogen detector, etc.

The other area in which the electron capture detector has tremendous use is in research activities where a known compound is studied under controlled conditions. In this kind of study, the substrate can be checked for interferences, fortified control samples can be used to determine response of the compound of interest in the presence of the substrate extractives, and levels in the treated samples then determined with a high degree of reliability, provided the work is done by a skilled residue analyst who will immediately recognize any irregularities.

Perhaps the most unfortunate property of the electron capture gas chromatograph is that almost anyone can take one and, after a few hours of instruction from a salesman, can inject samples and obtain recorder responses. This has happened, for there was an immediate rush to get on the band wagon by many workers in various phases of pesticide research and related fields who had absolutely no training in pesticide residue analysis or in the use of instruments such as the gas chromatograph. The result is that there is an unknown, but very substantial, amount of data in the literature that is of absolutely no value and which, more important, may be misleading. It was realized, belatedly by many, that electron capture responses are only indicators and cannot be accepted until confirmed by an accepted independent method. An outstanding illustration of misinterpreted electron capture gas chromatographic responses is found in the multitude of reports of DDT residues in every type of environment in every part of the world. It is only now becoming evident that many of these reported DDT residues were, in fact, polychlorinated biphenyls, and it may be found eventually that other compounds have been mistaken for DDT when suitable confirmatory procedures were not used.

The Microcoulometric Detector

One of the most useful detectors for gas chromatographic determination of pesticide residues, the microcoulometric detector, was developed by Coulson *et al.* (*3, 4*) and described by Coulson (*5*). This was the first system to be developed specifically for pesticide residue determination. Cassil (*6*) described the instrument and its application in the determination of Thiodan in the presence of DDT. Over the ensuing years, the microcoulometric detection system became virtually a necessity in any laboratory determining organochlorine or sulfur-containing organic pesticides. In addition to its use as a primary tool for analysis, the detector has very wide use as a confirmatory procedure for electron capture detector responses appearing to be organochlorine pesticides. The specificity for chlorine, or sulfur, as the case may be, allows the operator a high degree of confidence in the results obtained and has led to the extensive use of this detection system for confirmatory purposes by the U. S. Food and Drug Administration and other regulatory agencies.

During the past eight years, many improvements have been made in the detection system. An improved titration cell and a better electrometer and pyrolysis furnace increased the response many-fold and simplified the maintenance problems associated with the various components. The latest model, recently announced, is completely new, with the exception of the titration cell, and embodies a miniaturized pyrolysis furnace and improved coulometer and power supply as well as a completely new cabinet configuration. A newly-designed pyrolysis tube, it is claimed, has resulted in a system that is insensitive to changes in gas flow rates and gives increased response. Doubling or halving the oxygen flow rate does not appreciably affect response, nor do substantial changes in the carrier gas flow rate. This is in sharp contrast to a miniaturized pyrolysis system announced by Guiffrida and Ives (*7*) that gave substantial changes in response with changing gas flow rates.

Tests with the new production model show that the oxygen used for combustion must be saturated with water vapor to give maximum and consistent response. This was reported by Coulson (*5*) and by Barkley and Gunther (*8*) for the first model produced but has been largely overlooked by users of this equipment.

The latest production model of the microcoulometric detection system has a minimum detectable limit of about 1 ng of chlorine but, in the writer's opinion, the minimum for practical use is about 3 ng. This estimate is based upon responses obtained by Dohrmann Instruments personnel with standard solutions of pure lindane. These results were obtained in mode I operation (gas flow into the cell between the electrodes) rather than mode II operation (gas flow impinging directly on the sensor electrode), a more "sensitive" mode currently being used on last year's model detector in our laboratories. Mode I is preferred, for operating parameters are much less critical.

In 1966, Martin (*9*) announced a titration cell for the microcoulometric system that is specific for nitrogen after pyrolysis of the sample under reducing conditions during which the nitrogen is converted to ammonia. Albert (*10*) extended this work, and Cassil *et al.* (*11*) have more recently reported on the use of this detector for determining organonitrogen pesticides and have compared its performance with the electrolytic conductivity detector, finding them equal. These authors found both detectors useful in the range of 3 to 200 ng of nitrogen, permitting the measurement of as little as 0.05 ppm of organonitrogen pesticides in 100-gram samples.

Burchfield and his associates (*12, 13, 14*) reported a procedure involving reductive pyrolysis of the effluent gases whereby compounds containing chlorine, sulfur, or phosphorus form HCl, H_2S, or PH_3, respectively. By using an aluminum oxide scrubber following the pyrolysis stage, the system was specific for phosphorus. However, the detection system appears to be inferior to other available detectors for phosphorus and will probably have little or no use for this purpose.

The Electrolytic Conductivity Detector

The electrolytic conductivity detector for gas chromatography was developed by Coulson (*15, 16, 17*), who described modes of operation for the detection of chlorine, sulfur, or nitrogen, but did not establish the reliability of the detector for pesticide residue analysis or the minimum detectability for each molecular species. Cassil *et al.* (*11*) described the use of the detector for determining residues of carbamate pesticides and compared its response with that of the microcoulometric detector, as mentioned earlier, finding them equal in response and selectivity and usable over a range of 3 to 200 ng of nitrogen. An improved pyrolysis tube was described, and nickel wire or turnings was used as the catalyst instead of the nickel sulfate-treated quartz wool suggested by Coulson. Various alkalies were tested between the pyrolysis furnace and detector, barium oxide on 10/16 mesh perlite being preferred.

Patchett (*18*) made a detailed study and evaluation of the electrolytic conductivity detector and developed techniques to permit dependable and continuous use of the detector near a lower limit of detection of 0.1 ng of nitrogen. He made a substantial improvement in detector response through modifying the detector cell by placing a Teflon insert in the inlet tube to reduce sorptive losses of ammonia. He also proved the necessity for using high-purity hydrogen gas and maintaining high-purity deionized water between pH 7.0 and 7.5. Both Patchett and Cassil recommend the use of hydrogen as the carrier gas to obtain optimum consistent results.

Westlake *et al.* (*19*) used the electrolytic conductivity detector for determining residues of ACD15M, a triazine herbicide (Allied Chemical Co.) in sweet corn grain and plants (stalks and leaves). They also demonstrated the separation of a mixture of eight triazine herbicides and the absence of interferring responses at the retention time of the compound either by naturally-occurring plant components or related pesticides. A major peak, from an unidentified component of the corn plant extractives, was observed at a retention time about three times that of ACD15M. It was not present in the corn grain. Information to date indicates that such an interference is rare, but this serves to illustrate the necessity for vigilance at all times, even when using a detector as specific as this one, to guard against incorrect interpretations of the recorder responses. In general, the electrolytic conductivity detector is remarkably free from interferences from naturally-occurring components in plant extractives and anything other than organonitrogen compounds, but there are frequent exceptions.

The electrolytic conductivity detector has been used to determine organic iodine with excellent success by Westlake (*20*), as well as chlorine in organochlorine pesticides, operating in the reducing mode to yield HI or HCl as the detected product. The minimum detectability for chlorine is approximately equal to that of the current microcoulometric detection system. Coulson (*4, 15*) compared the responses of the electrolytic conductivity, microcoulometric, and electron capture detectors for organochlorine compounds in various extractives and found the first two approximately equal and the electron capture detector unsatisfactory because of high background.

One important advantage of the electrolytic conductivity detector is its simplicity. It requires no amplification of signal, thus eliminating the electrometer necessary for the other systems. In the reducing mode, it is highly specific for halide or, with alkali scrubber, for nitrogen. In the oxidative mode, it is quite specific for sulfur or chlorine.

Summary

Each of the detectors discussed has an extremely important place in pesticide residue determinations. When the limitations and capabilities of each are recognized and the responses evaluated by experienced personnel, a reasonable assurance of the identity of many pesticides can be obtained. None of the detectors are completely specific and, for positive identification, verification by an independent method is required.

Unfortunately, many investigators failed to realize (and this is still true of some) that the gas chromatograph can and does lie, and a tremendous volume of data of doubtful merit has been published. The more experienced one becomes with the technique, the more skeptical he becomes about the validity of the gas chromatographic responses. Only by maintaining this attitude can valid data be derived from gas chromatography of pesticide residues.

Literature Cited

(1) Westlake, W. E., Gunther, F. A., *Residue Rev.* (1967) **18**, 175.
(2) Lovelock, J. E., Lipsky, S. R., *J. Am. Chem. Soc.* (1960) **82**, 431.
(3) Coulson, D. M., Cavenaugh, L. A., *Anal. Chem.* (1960) **32**, 1245.
(4) Coulson, D. M., De Vries, J. E., Walther, B., *J. Agr. Food Chem.* (1960) **8**, 399.
(5) Coulson, D. M., *Advan. Pest Control Res.* (1962) **5**, 153.
(6) Cassil, C. C., *Residue Rev.* (1962) **1**, 37.
(7) Guiffrida, L., Ives, N. F., *J. Assoc. Offic. Agr. Chemists* (1969) **52**, 541.
(8) Barkley, J. H., Gunther, F. A., unpublished report, 1964.
(9) Martin, R. L., *Anal. Chem.* (1966) **38**, 1209.
(10) Albert, D. K., *Anal. Chem.* (1967) **39**, 1113.
(11) Cassil, C. C., Stanovick, R. P., Cook, R. F., *Residue Rev.* (1969) **26**, 63.
(12) Burchfield, H. P., Johnson, D. E., Rhoades, J. W., Wheeler, R. J., *J. Gas Chromatog.* (1965) **3**, 28.
(13) Burchfield, H. P., Rhoades, J. W., Wheeler, R. J., *J. Agr. Food Chem.* (1965) **13**, 511.
(14) Burchfield, H. P., Wheeler, R. J., *J. Assoc. Offic. Agr. Chemists* (1966) **49**, 651.
(15) Coulson, D. M., *J .Gas Chromatog.* (1965) **3**, 134.
(16) *Ibid.*, (1966) **4**, 285
(17) Coulson, D. M., "Advances in Chromatography," Vol. III, p. 197, Marcel Dekker, New York, 1966.
(18) Patchett, G. G., *J. Chromatog. Sci.* (1970) **8**, 155.
(19) Westlake, W. E., Westlake, A., Gunther, F. A., *J. Agr. Food Chem.* (1970) **18**, 685.
(20) Westlake, A., unpublished data.

RECEIVED June 12, 1970.

6

Infrared Microtechniques Useful for Identification of Pesticides at the Microgram Level

ROGER C. BLINN

American Cyanamid Co., Princeton, N. J. 08540

Infrared spectrophotometry has a long history of usefulness for identification of milligram amounts of organic compounds. In recent years microtechniques have been developed for extending this utility to microgram amounts of material. Factors involved in achieving this sensitivity and in minimizing the effects of contamination are discussed. Isolation of pesticides from the sample for infrared characterization is most conveniently achieved by thin-layer chromatography; however, trapping gas chromatographic effluents is also advantageous. Various procedures for preparing micro potassium bromide pellets are described, as well as the use of micro multiple internal reflectance.

The subject of "Identification of Pesticides at the Residue Level" is both timely and important in this period when we are actively reviewing the values and hazards of pesticide usage, since such a review is dependent upon reliable analytical data. Therefore, the reporting of unconfirmed residue data is misleading and can often result in controversy. The use of infrared spectrophotometry has pioneered in this important task of confirming the identity of pesticide residues.

Infrared spectrophotometry has a long history of usefulness in helping to establish and to confirm the identity of organic compounds. Functional group–absorption band correlation charts are well known and have been used routinely by organic synthesis chemists and by analysts for characterizing compounds of unknown identity. Where a synthetically prepared compound is not available for comparison with the unknown, infrared data in conjunction with mass, ultraviolet, and nuclear magnetic resonance spectral data can allow the deduction of the unknown compound's structure with reasonable assurance. However, positive identification must await exact comparison of the various properties of the unknown with those of a compound prepared synthetically by an unequivocal procedure. In practice, exact infrared spectral comparison alone usually constitutes confirmation of identity.

The low cost of many fine infrared spectrophotometers has contributed to their availability to most pesticide residue analysts, as contrasted to mass and nuclear magnetic resonance spectrometers. The sensitivity inherent in infrared measurements for identification purposes is only exceeded by mass spectrometry. This sensitivity has been realized by the development of suitable and practicable microtechniques, and some of these have been available for at least 10 years.

Now, what are the factors that can be varied in order to achieve sensitivity? The law describing the absorption of electromagnetic energy by absorbing substances in solution, Beer's Law, states that the absorbance of a solution is proportional to the absorptivity of the absorbing compound, the distance the energy beam passes through the solution, and the concentration of the absorbing compound in the solution. In other words, sensitivity can be achieved by placing the maximum number of the available molecules of a compound in the usable energy beam of the spectrophotometer. This can be accomplished by increasing the distance the energy beam passes through the solution, and/or by increasing the concentration of the sample in the solution by decreasing the volume requirements of the spectrophotometer's sample holding accessory. The word "solution" may also refer to the essentially pure compound, either as a film or as particulated very finely in a medium.

The other factor in Beer's Law affecting sensitivity is the absorptivity of the sample. The analyst, of course, has no control over this factor, but he certainly should be aware of it. He should know that a larger sample of the low-absorptivity cyclodiene pesticides, for example, will be required to achieve a satisfactory spectrum than for the more strongly-absorbing organophosphorus pesticides.

Of course, the real challenge to achieving sensitivity is in eliminating or minimizing the ever-present sources of interference. As the size of the sample decreases into the microgram range, decreasing the interferences at the same rate is increasingly difficult. The sources of interference are everywhere, arising from the sample, solvents, reagents, atmosphere, handling, and other similar sources. Many of these interferences can be minimized by due care in choosing isolation procedures, pre-washing sorbents and glassware, purifying solvents and reagents, and proper manipulative techniques. But complete elimination of all interfering materials is extremely difficult to accomplish. Thus, interferences do limit the sensitivity that can be achieved. Chen and Dority (*1*) recently discussed the importance of minimizing interferences in microsampling.

What order of sensitivity is achievable in the infrared with the techniques and equipment presently available? Several workers have proposed that the absolute lower limit of sensitivity is 10 ng of a compound with moderately strong absorptivity values (*2, 3*). However, because of the difficulty in placing all or most of the sample in the actual usable energy beam of the spectrophotometer, an actual lower limit of sensitivity is about one-tenth to one-half of a microgram, that is, about 100 to 500 ng (*4, 5*). For most workers, this limit will be about five to ten micrograms. One does not have to be a professional infrared spectroscopist to attain success in the microgram range. Students that have never previously used an infrared spectrometer have achieved excellent spectra with ten micrograms or less of a pesticide after a few hours of instruction and practice, so microtechniques in the infrared are usable by the pesticide residue analyst.

As stated previously, the pesticide must be isolated from the sample, whether the sample is natural waters, soil, or plant or animal tissue. This isolation must be virtually complete. Usually, a chromatographic process is used to achieve this separation of the pesticide from natural substances extracted from the sample and from other pesticides and foreign substances residing in the sample. Vapor phase and thin-layer chromatography have traditionally been the methods of choice for this purpose. When using either of these chromatographic processes in order to isolate a pesticide for infrared scrutiny, it is advantageous to subject the sample extract to a rigorous "cleanup" prior to chromatography. With a decreased amount of extraneous material to be separated from the pesticide, the chromatographic process will be more efficient. Even with a rigorous cleanup, a sample may require several chromatographic isolations before reliable data can be obtained. However, for each process to which the sample is subjected, there is an inevitable loss of a portion of the sample. The loss may be minor, but a portion of the sample is lost from the infrared procedure. Therefore, the starting sample must be of sufficient size so that there will be enough pesticide isolated *via* all of the isolation procedures to allow infrared evaluation.

The well-deserved popularity of gas chromatography in the field of pesticide residue analysis suggests that this tool be used for isolation purposes. The trapping of gas chromatographic peaks for infrared identification has been and is being used. Several factors must be considered in the successful use of gas chromatography for trapping and the meaningful evaluation of the spectrum of the trapped material. The high temperatures and the large, often catalytically active, surface areas that are encountered in the gas chromatographic process can change the chemical nature of the sample. Gas chromatography of such compounds can result in usable and reliable elution peaks for measurements, but could be misleading for identification purposes. The larger amounts of material needed for infrared purposes compared with analytical studies makes the chromatographic process more difficult and increases the possibility for chemical alteration of the pesticide. Another factor to be considered when trapping a gas chromatographic peak is that the specificity of a detector will often obscure the simultaneous elution of interfering substances from the sample and the column. Such detectors as the electron capture, thermionic, flame photometric, microcoulometric, and nitrogen detectors respond selectively to certain types of compounds and insensitively or not at all to others. This is the reason that these detectors were chosen for pesticide residue analysis. Therefore, a detector response as a sharp peak from a baseline may either be caused by the elution of a single compound or by this compound in company with other material to which the detector is unresponsive. The latter case could be unsuitable for infrared purposes. Another source of interference of this type is from column "bleed." The stationary phase of any gas chromatographic column does possess a certain amount of volatility and will slowly elute and often collect at the exit port. The investigator should be familiar with the infrared spectrum of his stationary phase. Figure 1 shows the spectrum of one of the column materials used in pesticide analysis, silicone oil. Everyone using infrared techniques should know this spectrum, as it can arise as well from stopcock gease on glassware (*1*).

Because of the destructive nature and/or extreme sensitivity of commonly used detectors for pesticide problems, the gas chromatograph used for pesticide analysis usually is not suitable for trapping. Although splitters are available that can be used to convert the analytical instru-

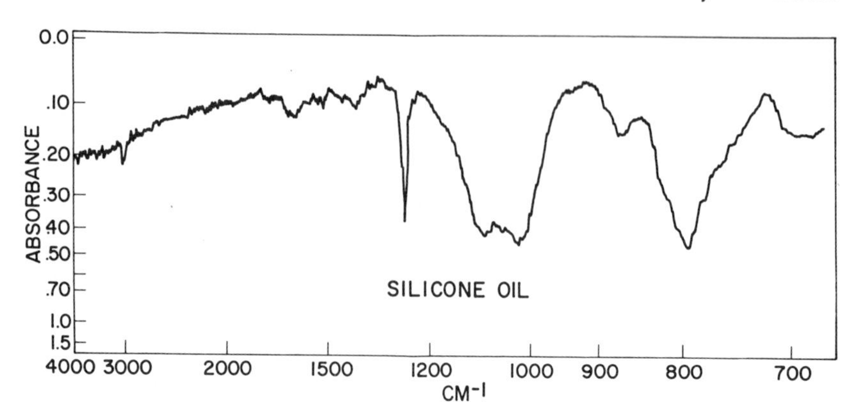

Figure 1. Infrared spectrum of silicone oil

ment for preparative purposes, this usually is not satisfactory owing to the lost instrument time for analytical purposes and the man-hours and frustration required to achieve conversion. A preparative instrument which is used only for identification purposes is preferable.

The actual trapping procedure to be used is a matter of individual preference, since there are many published procedures and commercial types available. The capillary-type trapping device (*6, 7, 8*) is appealingly simple, and the concentrated trapped pesticide is confined on the small inner surfaces of the tube from which it is quickly, easily, and, more important, efficiently transferred to whichever infrared microsampling device is to be used.

Thin-layer chromatography is a more adaptable procedure for the isolation of microgram quantities of a pesticide prior to its identification by infrared evaluation, and the thin-layer process aids in the identification as well. Locating the area in which the pesticide resides on the developed chromatogram can be a problem, however. A sorbent with fluorescent indicator can be used for those compounds which quench fluorescence. Alternatively, the chromatography of standards in a side channel can be used for locating purposes or an aliquot of the unknown solution can be chromatographed in a side channel for colorimetric detection. Once located, the sorbent from that area can be scraped from the plate for elution. Elution should be accomplished with a minimum amount of as nonpolar a solvent as possible, thus restricting the amount of co-eluting interferences from the sorbent. Very polar solvents will elute interferences from the sorbent which are very difficult to cleanup (*1, 4*). The sorbent on the thin-layer plate should have been pre-washed with the eluting solvent or a solvent of greater polarity prior to the chromatography, of course. One should not overlook the possibility for chemical alteration of the pesticide on the sorbent, as several types of compounds are subject to such changes—*e.g.*, phenols and amines. Also, very polar compounds are often difficult or impossible to elute from the sorbent. Aside from these limitations, thin-layer chromatography is advantageous for isolating pesticides prior to infrared evaluation. The simple equipment and techniques required and the relative freedom from chemical alteration makes this approach useful to all.

Now as to the microtechniques which have proved useful for infrared spectrometry, there are several types available in the literature, such as microcells for solutions (*9, 10*), microgrooved plates for confined films (*7, 11*), suspended particles on a membrane filter (*12, 13*), microspecular spectrometry (*14*), the various micropellet procedures (*15, 16, 17, 18, 19, 20, 21, 22, 23, 24, 25*), and multiple internal reflectance (*5, 12, 15, 26, 27*). Then, of course, there is multiple scan interference spectrometry (*28*) with computer storage and handling of the data. It is with this technique that significant gains in sensitivity and speed would appear promising for the future. But for the present, the use of micro potassium bromide pellets and multiple internal reflectance are proving most useful for pesticide problems.

Of all of the microsampling techniques described for infrared spectrometry, the use of potassium bromide pellets of 1–2 mm in diameter offers the best opportunity for placing the maximum number of molecules of the unknown in the usable energy beam of the spectrophotometer. Certainly, the best sensitivities have been realized by this technique. The equipment commonly used for preparing micro potassium bromide pellets is shown in Figure 2. The key to sensitivity is the ability to transfer the maximum amount of compound to the minimum amount of potassium bromide to be pressed into the micropellet. Conventional mixing procedures using a mortar and pestle, such as is shown in Figure 2, are subject to very large losses of compound to the surface area of the mixing vessel when only 10 mg or less of potassium bromide are used (*1, 18*). One of the first attempts to resolve this problem was the procedure illustrated in Figure 3, which was developed by McCaulley (*21*) and then refined by Chen (*16*). With this technique, powdered potassium bromide was packed tightly in the 1.5-mm orifice of the stainless steel disk. The sample in solution of a volatile solvent was added dropwise to the potassium bromide, allowing the solvent to evaporate between additions. After completing the additions, the pellet was pressed in the conventional manner for infrared evaluation. This procedure theoretically offers complete transfer of the sample to the energy beam of the spectro-

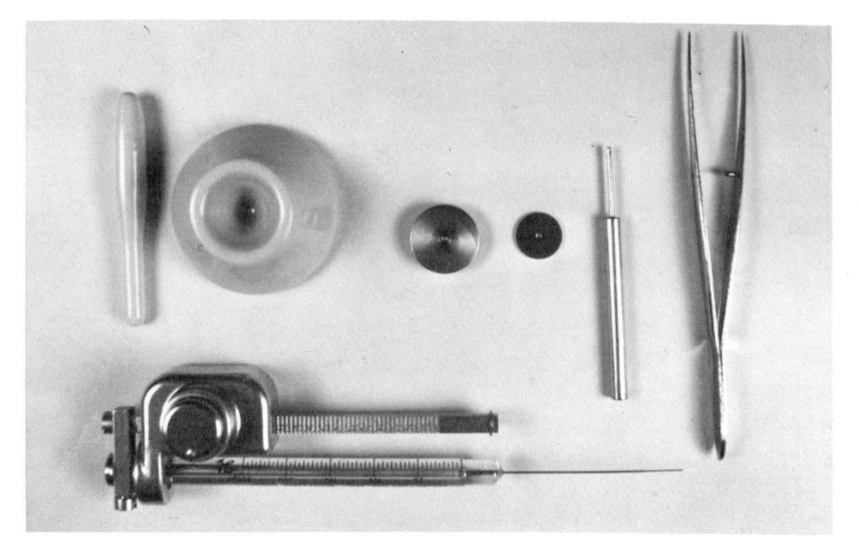

Figure 2. Equipment useful in preparing micro potassium bromide pellets

Figure 3. Illustration of technique described by McCaulley (21) *and Chen* (16)

photometer. Practically, this usually does not happen since a portion of the solution is preferentially drawn from the potassium bromide because of the greater surface tension of the metal disk. The losses are often very great, but with due care and steady nerves, excellent sensitivity can be realized with this technique.

Another technique developed to minimize losses of compound on container surfaces is described by Chen and Dority (*1*) and by de Klein (*29*), in which the sample is deposited in a capillary tube. The powdered potassium bromide can be added to the capillary tube either prior to the addition of the sample or afterwards; mixing is accomplished in the latter case by shaking and letting the scouring action of the potassium bromide reduce residual tubing surface losses.

Curry *et al.* (*18*) describe another technique, shown in Figure 4, in which they "dispense" about half of a microliter of a chloroform solution of the sample to the tip of a syringe needle, "picking up" powdered potassium bromide by adhering action on the needle tip, dispensing another aliquot of the solution to the powder, and evaporating the chloroform. The "dispensing" of the solution is continued until the entire sample is transferred to the powder, which is then pressed into a micropellet. This procedure is theoretically sound but would seem to be nerve-wracking in operation. Actually, the greatest difficulty is in achieving the initial adhesion of the powdered potassium bromide to the syringe needle. Once adhered, the powder will stick to the needle tip until completely dry of solvent.

Two techniques for transferring samples to potassium bromide powder are especially suited for use with thin-layer chromatography. A procedure described by de Klein (*30*) is illustrated in Figure 5. It in-

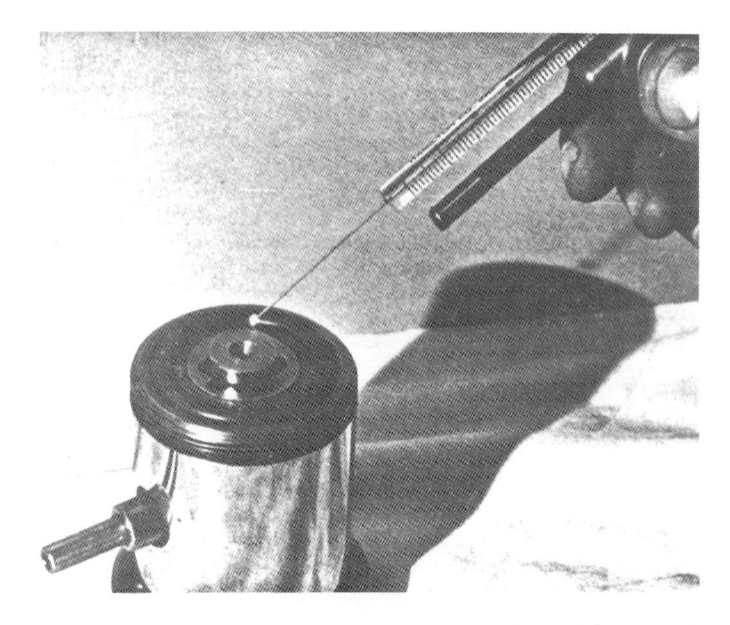

Journal of Chromatography

Figure 4. Illustration of technique described by Curry et al. (18)

volves scraping the sorbent from around the chromatographic spot of interest, placing a "dam" of powdered potassium bromide around the tip of the spot, eluting the compound with a volatile solvent, and allowing the solvent to spill over to the "dam" of potassium bromide where it is allowed to dry. The other technique involves using the commercially available "wick stick" and is described by Krohner and Kemmner (5). The wedge of potassium bromide is dipped at its base into a solution of the compound in a volatile solvent. The solution migrates up the wedge to the tip, where it evaporates, and the compound is concentrated in the tip. The tip is then broken from the wedge and pressed into a micropellet. The use of "wick sticks" is simple, reliable, and recommended for preparing micro potassium bromide pellets.

All of these microtechniques are successful in transferring the sample to the potassium bromide micropellet, but losses of sample are inherent for each of them. These losses can be very great without attention to detail. Elimination of water absorption bands is also quite difficult to accomplish. A further disadvantage of this procedure is the difficulty in recovery of the sample for evaluation by other means. But the excellent sensitivity achievable does recommend this sampling procedure for micro quantities of material.

In Figure 6 is presented the principle by which infrared spectra are obtained by multiple internal reflectance. The infrared energy is

Analytical Chemistry

Figure 5. Illustration of technique described by de Klein (30)

directed into the entrance face of the infrared transparent crystalline plate. This energy is reflected repeatedly from the inner faces of the top and bottom surfaces as it transverses the plate, emerges through the exit face, and is directed into the spectrophotometer. The sample is plated as a film on the top and bottom surfaces of the plate. At each totally internal reflection there is a penetration of the electromagnetic field into the rarer medium beyond the reflecting interface. Any sample in contact with the surface is penetrated a few microns, and a spectrum results. The most useful crystalline material for multiple internal reflectance spectrometry is KRS-5, a mixture of thallous bromide and thallous iodide. When using a 50- × 20- × 1-mm plate, sensitivities of about 15 to 20 μg are routinely achieved with stronger absorbing com-

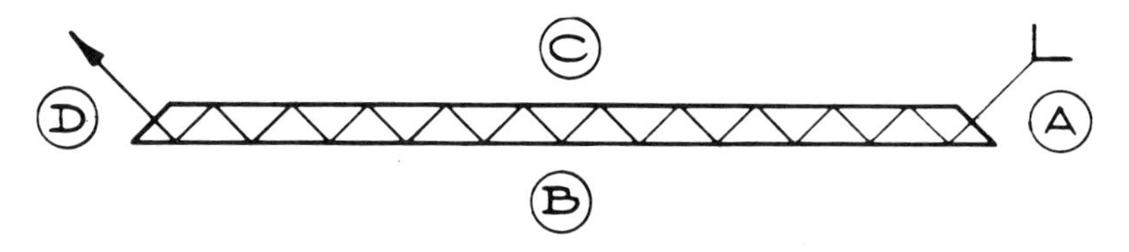

Figure 6. Pathway of infrared energy beam in multiple internal reflectance plate

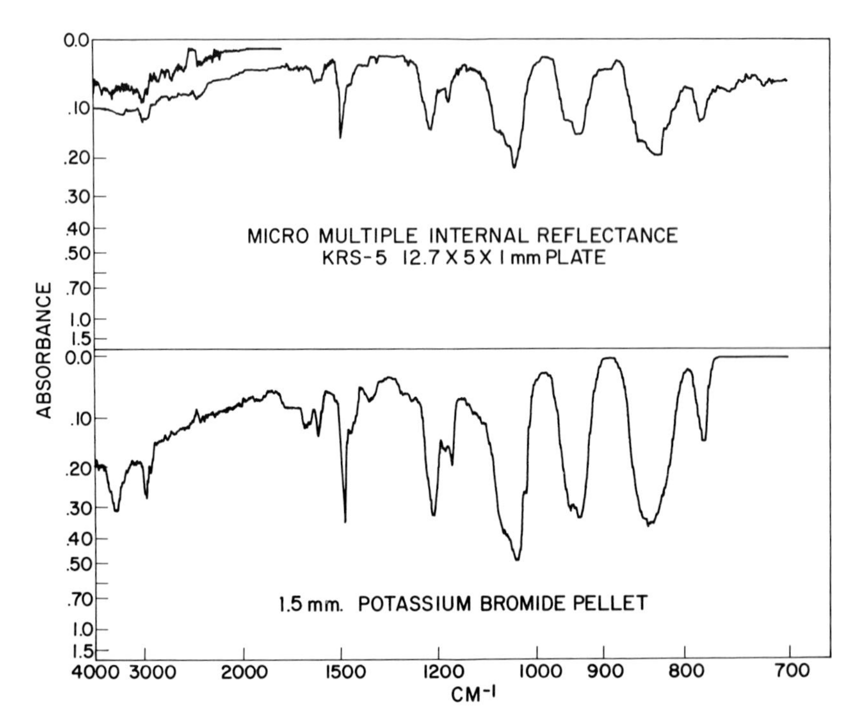

Figure 7. Infrared spectra of Abate mosquito larvicide by multiple internal reflectance and micro potassium bromide pellet

pounds. There is now a micro version of this technique introduced by Wilks Scientific Co. using a 12.7- × 5- × 1-mm plate which will allow sensitivities of about 3 to 5 μg. The great advantage of multiple internal reflectance is the ease of applying the sample to the surface of the reflectance plate. It is just streaked or dotted as a solution onto the surfaces, allowing the solvent to evaporate. After the spectrum is recorded, the sample can be washed from the surfaces for any further evaluation desired. The only real difficulty with sample preparation is with those compounds whose crystalline structure makes intimate contact with the surface difficult and those materials which react with the thallous bromide or iodide. The spectrum resulting from multiple internal reflectance is similar but slightly different from transmission spectra, since absorption is greater at the longer wavelengths.

Figure 7 shows the spectra obtained from 4.9 μg of Abate mosquito larvicide, using both the micromultiple internal reflectance and micro potassium bromide pellet techniques. The technique of Curry *et al.* (*18*), dispensing the solution of the insecticide onto the potassium bromide powder adhered to a syringe needle, was used. In Figure 8 is shown the spectra of 9.8 μg of technical DDT obtained in the same manner, and Figure 9 shows similarly prepared spectra for 24.6 μg of endrin. These three figures were prepared to illustrate the sensitivities achievable, the superior sensitivity of the micro potassium bromide pellet technique, and the greater sensitivity achieved for the highly absorbing organophosphorus pesticide than for the organochlorine compounds.

In conclusion, three factors are very important to microtechniques in infrared spectrophotometry. First, in order to gain sensitivity, an efficient transfer of the sample to the usable energy beam of the spectrophotometer must be achieved. Microtechniques strictly limit the amount of allowable contamination from such sources as the sample, solvents, sorbents, reagents, atmosphere, handling, and the like.

Secondly, start the isolation procedure with sufficient sample so that there will be enough of the finally isolated pesticide to allow a spectrum. Remember that each handling step in the isolation procedures and the microinfrared techniques results in some loss of the sought pesticide.

Third and last, infrared spectrophotometry is only one tool used by the pesticide residue analyst, and he cannot be expected to become

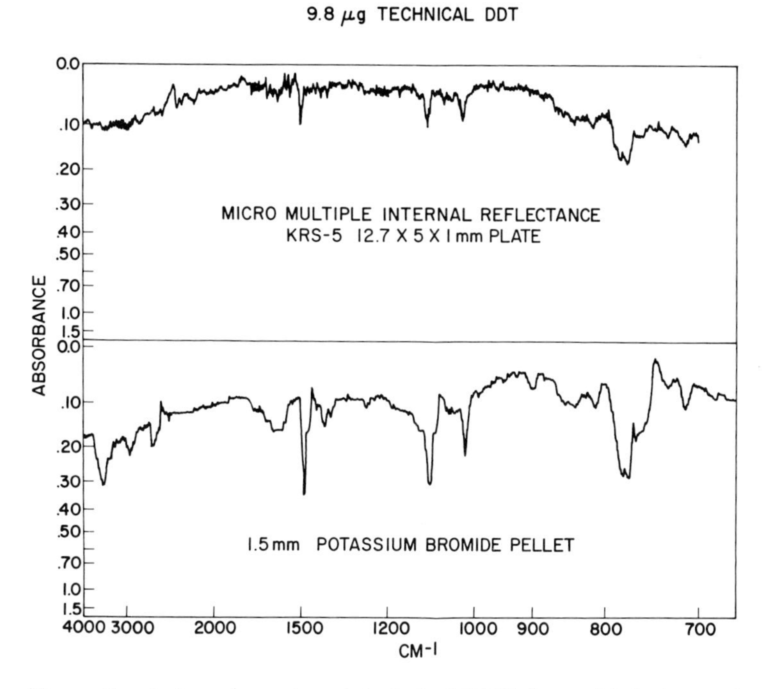

Figure 8. Infrared spectra of technical DDT by multiple internal reflectance and micro potassium bromide pellet

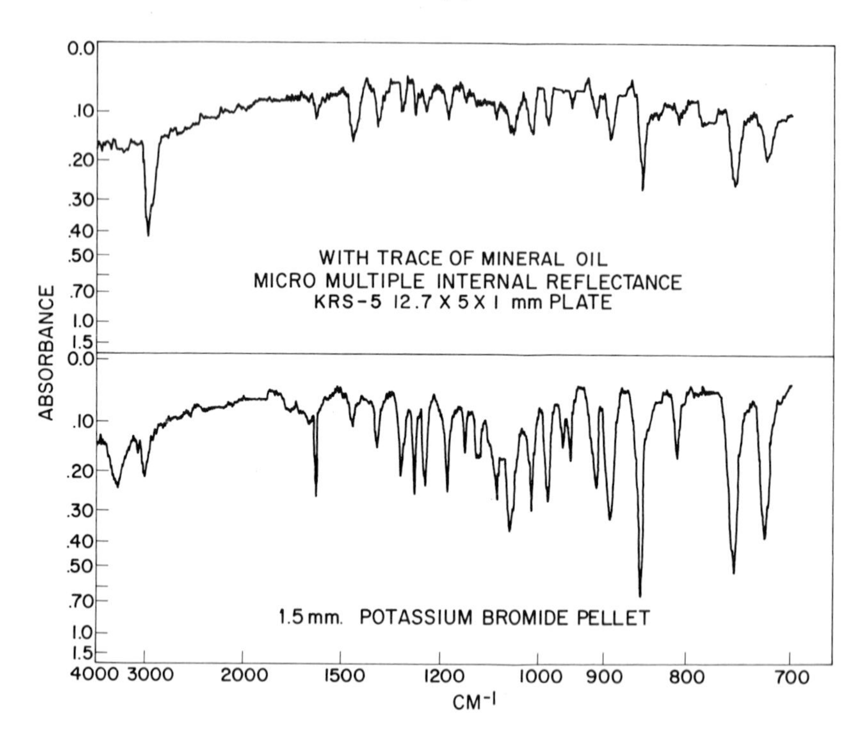

Figure 9. Infrared spectra of endrin by multiple internal reflectance and micro potassium bromide pellet

as skilled as the professional spectroscopist. But he can achieve successful results in the 5–10 μg range if he uses the simplest techniques commensurate with the objectives of the problem.

Literature Cited

(1) Chen, J. Y. T., Dority, R. W., "Contamination Control in Infrared Microanalysis," *Annual Meeting of the Association of Official Analytical Chemists, 83rd, Washington, D. C.*, October 14, 1969.
(2) Mason, W. B., "Infrared Microspectrophotometry," *Microchem. J. (Symp. Ser.)* (1961) **1**, 293–310.
(3) Traber, W. F., "'Thinking Small' with Microanalysis Techniques," *Ind. Res.* (October 1966), pp. 80–85.
(4) Fahr, E., Rohlfing, W., "Infrared Spectroscopy in Biochemistry and Clinical Chemistry," *Z. Anal. Chem.* (1968) **243**, 43–48.
(5) Krohner, P., Kemmner, G., "Methods and Detection Limits in the IR-Spectroscopic Investigation of Micro Quantities," *Z. Anal. Chem.* (1968) **243**, 80–92.
(6) Edwards, R. A., Fagerson, I. S., "Collection of Gas Chromatographic Fractions for Infrared Analysis," *Anal. Chem.* (1965) **37**, 1630.

(7) Kabot, F., "How to Collect Fractions from the Model 800 Series Gas Chromatograph," Perkin-Elmer Corp., *G. C. Newsletter* (1967) **3 (1),** 1–4.
(8) Oadland, R. K., Glock, E., Bodenhamer, N. L., "A Simple Technique for Trapping Gas Chromatographic Samples from a Capillary Column for Mass Spectrometry or Re-Chromatography on Another Column," *J. Chromatog. Sci.* (1969) **7,** 187–189.
(9) Crosby, N. T., Laws, E. Q., "The Use of Infrared Spectroscopy in the Analysis of Pesticide Residues," *Analyst* (1964) **89,** 319–327.
(10) Price, G. C., Sunas, E. C., Williams, J. F., "Micro Cell for Obtaining Normal Contrast Infrared Solution Spectra at the Five Microgram Level," *Anal. Chem.* (1967) **39,** 138–140.
(11) Mills, A. L., "Infrared Identification of Microgram Quantities of Heroin Hydrochloride," *Anal. Chem.* (1963) **35,** 416.
(12) Hannah, R. W., Dwyer, J. L., "Analysis of Suspended Particulates with Membrane Filters and Attenuated Total Reflection," *Anal. Chem.* (1964) **36,** 2341–2344.
(13) Sloane, H. J., "Infrared Differential Technique Employing Membrane Filters," *Anal. Chem.* (1963) **35,** 1556–1558.
(14) Sloane, H. J., Johns, T., Cadman, W. J., Ulrich, W. R., "Infrared Examination of Microsamples. Application of a Specular Reflectance System," *Appl. Spectry.* (1965) **19,** 130–135.
(15) Blinn, R. C., "Infrared Techniques Useful in Residue Chemistry," *J. Assoc. Offic. Agr. Chemists* (1965) **48,** 1009–1017.
(16) Chen, J. Y. T., "Micro-KBr Technique of Infrared Spectrometry," *J. Assoc. Offic. Agr. Chemists* (1965) **48,** 380–384.
(17) Chen, J. Y. T., Gould, J. H., "Micro-AgCl Technique of Infrared Spectrometry," *Appl. Spectry.* (1968) **22,** 5–7.
(18) Curry, A. S., Read, J. F., Brown, C., Jenkins, R. R., "Micro Infrared Spectroscopy of Gas Chromagraphic Fractions," *J. Chromatog.* (1968) **38,** 200–208.
(19) Garner, H. R., Packer, H., "New Techniques for the Preparation of KBr Pellets from Micro-Samples," *Appl. Spectry.* (1968) **22,** 122–123.
(20) Hayden, A. L., Brannen, W. L., Craig, N. R., "A Micro-Extraction Technique with Compounds Isolated from Thin-Layer Chromatograms," *J. Pharm. Sci.* (1968) **57,** 858–860.
(21) McCaulley, D. F., "An Approach to Separation, Identification, and Determination of at Least Ten Organophosphate Pesticide Residues in Raw Agricultural Products," *J. Assoc. Offic. Agr. Chemists* (1965) **48,** 659–665.
(22) Mount, D. I., Boyle, H. W., "Parathion—Use of Blood Concentrations to Diagnose Mortality of Fish," *Environ. Sci. Technol.* (1969) **3,** 1183–1185.
(23) Payne, W. R., Jr., Cox, W. S., "Micro-Infrared Analysis of Dieldrin, Endrin, and other Chlorinated Pesticide Residue in Complex Substrates," *J. Assoc. Offic. Agr. Chemists* (1966) **49,** 989–996.
(24) Robbins, J. D., Bakke, J. E., Fjelstul, C. E., "Practical Micro-KBr Disk Techniques for Infrared Spectrometry," *157th Meeting, ACS, Minneapolis, Minn., April 14, 1969.*
(25) Sterling, K. J., "Preparation of Potassium Bromide Disks for Infrared Microanalysis by Using a Half-Inch Die," *Anal. Chem.* (1966) **38,** 1804.
(26) Hermann, T. S., "Identification of Trace Amounts of Organophosphorus Pesticides by Frustrated Multiple Internal Reflectance Spectroscopy," *Appl. Spectry.* (1965) **19,** 10–14.

(27) Wilks Scientific Corp. (South Norwalk, Conn.), "Internal Reflectance Spectrometry," Vol. 1, 40 pp., 1965.
(28) Low, M. J. D., "Application of Multiple-Scan Interferometry to the Measurement of Infrared Spectra," *Appl. Spectry.* (1968) **22**, 463–471.
(29) de Klein, W. J., "Infrared-Spectroscopic Identification of Compounds Separated by Gas Chromatography, using a Potassium Bromide Micro-pellet Technique," *Z. Anal. Chem.* (1969) **246**, 294–297.
(30) de Klein, W. J., "Infrared Spectra of Compounds Separated by Thin-Layer Chromatography using a Potassium Bromide Micro-pellet Technique," *Anal. Chem.* (1969) **41**, 667–668.

RECEIVED June 29, 1970.

7

Ultraviolet Spectrophotometry in Residue Analysis; Spectra–Structure Correlations

OSMAN M. ALY and S. D. FAUST

Department of Environmental Sciences, Rutgers University, New Brunswick, N. J. 08903

I. H. SUFFET

Drexel Institute of Technology, Philadelphia, Pa.

Ultraviolet spectrophotometry is considered a valuable tool as an aid for confirming the identification of pesticide residues. A correlation between the UV spectrum and the structure of several pesticides is discussed. Knowledge of such correlation may provide clues about the general type of chromophore present and may help the analyst to design analytical procedures. The transparency of many groups in the near UV imposes a limitation on interpretations of the absorption bands in this region. However, when taken in conjunction with the information obtained by IR, NMR, and mass spectroscopy, UV spectra may lead to structural proposals of value to the pesticide analyst. A discussion of the methods that have been utilized for the analysis of pesticides on the submicrogram level is also presented.

Ultraviolet spectrophotometry has been, for some time, one of the most valuable tools of pesticide residue analysis. Any compound with a sufficiently well-defined and intense absorption spectra is potentially detectable by this spectrophotometric means. In some cases, suitable chromophores may be produced by chemical transformations. Blinn and Gunther (*1*) have reviewed the different procedures that have been utilized for residue analysis at the microgram level. UV spectrophotometry is considered one of the valuable aids for identification of organic pesticides. Although the ultraviolet data may not provide the multitude of information that can be gained from other spectrophotometric methods such as IR, NMR, or mass spectrometry, it is often possible to reveal subtleties of structure which these techniques cannot reveal (*2, 3, 4*).

During the past decade, tremendous advances have been achieved in the development of empirical correlations between the structure of organic molecules and their ultraviolet spectra (*4, 5*). This paper discusses the correlation between the spectra of some pesticides in the near ultraviolet region (190 to 400 μ) and their structure. Such information would be of value to the residue chemist for qualitative and quantitative purposes. Also, the application of UV spectrophotometry for residue analysis on the submicrogram level is presented.

Electronic Transitions

Absorption of quanta of radiation in the ultraviolet region results in a spectral transition in which the electrons of molecules are excited from their ground state to higher energy levels. The energy of a particle which is confined to a very small region is restricted to certain values. These energies are subjected to quantum restrictions which allow the molecule (or electrons) to have only certain energies. Generally, these restrictions are increasingly important as the region in which the particle is free to move becomes smaller. That is, of the whole range of energies, fewer and more widely spaced ones are allowed as the particle motion is more restricted. Therefore, the electrons which are confined to the volume, or perhaps, to a part of the volume of the molecule are subject to significant quantum restrictions or will be quantized. These restrictions allow the properties of molecules to be studied by the methods of spectroscopy (*6*). The energy of a quantum of radiation is calculated from Plank's equation

$$\Delta E = Kv \tag{1}$$

where ΔE is the energy of a quantum (ergs), K is Plank's constant (6.62×10^{-27} erg-sec), v is the frequency of wave motion in cycles/second, and

$$v = \frac{c}{\lambda} \tag{2}$$

where c is the velocity of light (3×10^{10} cm/sec) and λ is the wavelength in cm.

The energy of a quantum of radiation having wavelengths of 700, 400, 200, and 150 is, therefore, 2.85×10^{-12}, 4.97×10^{-12}, 9.9×10^{-12}, and 13.24×10^{-12} ergs, respectively.

When a molecule is irradiated by heterochromatic light, it absorbs only the protons whose energies are equal to that required for permissible energy transitions within the molecule. For most large molecules, the electronic absorption band is very complicated, and a rather broad absorption band with little or no detail is obtained. This is particularly true when the material is studied in solution.

Although absorption of ultraviolet light results in the excitation of electrons from their ground state, the nuclei which the electrons hold together determine the strength of the binding. Thus, the characteristic energy of transition, and hence the wavelength of absorption, is related to the electronic structure of a group of atoms within the molecule. The group producing absorption is called a chromophore. Structural changes affecting a chromophore can be expected to modify its absorption.

The three major contributors to electronic spectra are the single bond (σ-electrons), the multiple bond (π-electrons), and the unshared electron pair (n-electrons).

Compounds having only σ-valency electrons are saturated and do not absorb in the near ultraviolet region. All the electrons of the molecule are involved in single bonds, and they cannot be rearranged to an excited state without disruption of the molecular bonding. Excitation of these electrons requires high-energy photons in the far ultraviolet region ($\lambda <$ 150 μ). The transparency of the saturated hydrocarbons to 190 μ makes them excellent solvents for near ultraviolet spectroscopy.

The π-electrons are associated with compounds having double or triple bonds. Of particular interest to the residue analyst are these unsaturated compounds. The simplest compound in this group is the olefin ethylene ($CH_2{=}CH_2$). Three electrons of each carbon atom form shared pairs with the single electrons of the three neighboring atoms in much the same way as a single-bonded molecule. These shared pairs of electrons lie in a plane and the remaining electron of each carbon atom occupies the general region in space above and below this plane. Sharing of these two electrons (π-electrons), one from each carbon atom, makes the double bond (π-bond). These π electrons can be excited to different arrangements without disruption of the molecule. This transition is described as $\pi \rightarrow \pi^*$. In conjugated systems where double bonds and single bonds occur alternately, the π electron of one carbon atom can be shared with one of either neighboring carbon atom, resulting in delocalization of the π electrons. In these systems, the quantum restrictions become less important; *i.e.*, the allowed energies are more closely spaced the larger the region in which the particle can move. Therefore, ethylene absorbs at $\lambda = 175\ \mu$ ($\Delta E = 11.4 \times 10^{-12}$ erg) while butadiene absorbs at $\lambda = 210\ \mu$ ($\Delta E = 9.5 \times 10^{-12}$ erg). The delocalization effect is much more pronounced as the number of double bonds increases. The allowed energy levels become more closely spaced and the absorption moves from the ultraviolet region to the visible region—*i.e.*, the region of smaller quantum energies.

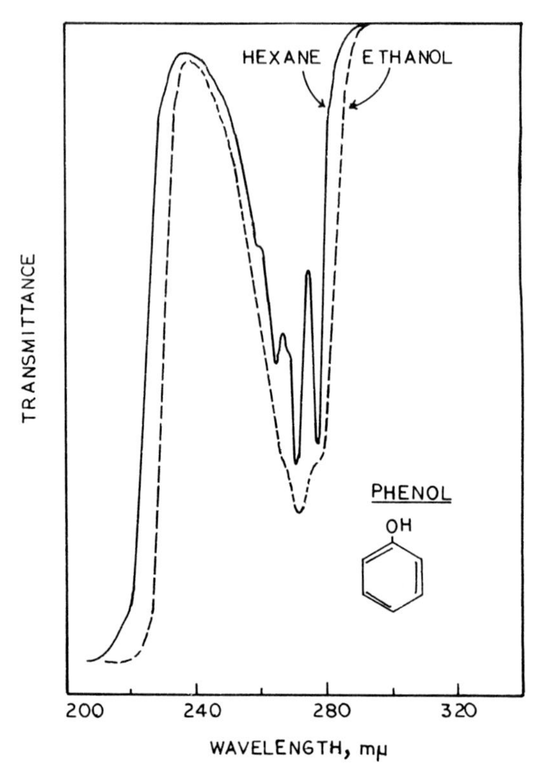

Figure 1. Ultraviolet absorption curve of phenol in hexane and ethanol

The *n*-electrons are those associated with hetero atoms such as N, O, S, and the halogens that are not involved in bonding (also called electron lone pairs). These electrons are held more loosely than the σ-electrons and undergo transitions at correspondingly higher wavelengths. This transition is designated $n \rightarrow \sigma^*$. The first absorption maximum of saturated compounds containing a hetero atom occurs in the far ultraviolet region and, therefore, they are not useful for diagnostic purposes. In addition, the relatively weak bands associated with their transitions are often obscured by other, more active, absorbing groups present in the molecule. The transparency of the chromophore $\geq$C—O—R or $\geq$C—Hal or $\geq$C—N— in the ultraviolet region explains the wide use of such compounds as ether, ethanol, and alkyl halides as spectroscopic solvents. In the alkyl halides, the bathochromic shift increases with increased substitution. Thus, the λ_{max} values of CH_3Cl, CH_2Cl_2, $CHCl_3$, and CCl_4 are 173, 216, 229, and 236 μ, respectively. The bathochromic shift increases with decrease of the electronegativity of the halogen atom (methyl iodide has a λ_{max} of 259). Compounds having a hetero atom involved in a double bond with a carbon atom such as C=O, C=N, and C=S exhibit bathochromic shifts. Two transitions are possible ($n \rightarrow \pi^*$ and $n \rightarrow \sigma^*$) as a result of absorption of radiation (*4*). Both the π electrons and the *n*-electrons are excited to a higher energy level. Acetone, for example, shows two absorption bands, one at 188 μ indicating $n \rightarrow \sigma^*$ transition and the other at 279 μ indicating $n \rightarrow \pi^*$.

Choice of Solvents in Spectral Analysis

The choice of a solvent is very important in studying absorption spectra of various compounds (*4, 7, 8, 9*). Spectra measured in saturated hydrocarbons tend to reveal more fine structures, and sometimes it is possible to observe the effects of intramolecular interactions more clearly. The fine structures revealed in these solvents illustrate the principle that nonsolvating or nonchelating solvents produce a spectrum near that obtained in the gaseous state. The use of hydroxylic solvents tends to smooth out the fine structures through solvent–solute interactions. Figure 1 shows the loss of fine structure in the spectrum of phenol in ethanol. The broad band owing to H-bonded solvent–solute complexes replacing the fine structure present in hexane is quite typical (*4*). Common solvents and their window regions are shown in Table I.

Major Chromophores in Organic Pesticides

Some of the chromophores that are repeatedly encountered in pesticides will now be discussed. The spectra presented here were measured in 95% ethanol and a 1-cm pathlength in a Beckman DB spectrophotometer. Wavelength values are accurate only to ± 2 μ. The transmit-

Table I. Minimum Wavelength for Common Solvents Used In UV Spectroscopy[a]

Solvent	*Wavelength* (μ) *of Cut-out*	
	1–mm Cells	10–mm Cells
Cyclohexane	190	195
Hexane	187	201
Carbon tetrachloride	245	257
Chloroform	223	237
Methylene chloride	215	220
Ethanol	198	204
Methanol	198	203
Water	187	191

[a] Taken from Scott, Ref. *4*, "Interpretation of the Ultraviolet Spectra of Natural Products," copyright by the Macmillan Company, 1964.

tancy values have been deliberately omitted and the wavelength scale is only presented for comparative purposes.

Aromatic Compounds. The correlation between the absorption spectra and structural features of aromatic compounds will be discussed because of the widespread occurrence of aromatic rings among pesticides. Although the infrared spectra form the most generally applicable method of recognition of the presence of a C–aromatic ring, valuable information may be gained by careful study of the ultraviolet spectra of certain classes of aromatic compounds.

BENZOID CHROMOPHORE. The spectrum of the aromatic parent, benzene (Figure 2), displays considerable fine structure, a property which is not shared to the same extent with many of its derivatives. The three-bonded spectrum (248, 254, and 260 μ) of benzene will be considered as one chromophore. Benzene absorbs at 184 μ (a_m 60,000), 203.5 μ (a_m 7400), and 254 μ (a_m 204) in hexane (*4*). These maxima are considered as the $\pi \rightarrow \pi^*$ bands of the benzene chromophore. Increasing alkyl substitution causes a bathochromic shift of the 254-μ band, an effect which reaches its maximum at tetrasubstitution. New intense bands appear in the spectrum of benzoid compounds upon introduction of a substituent

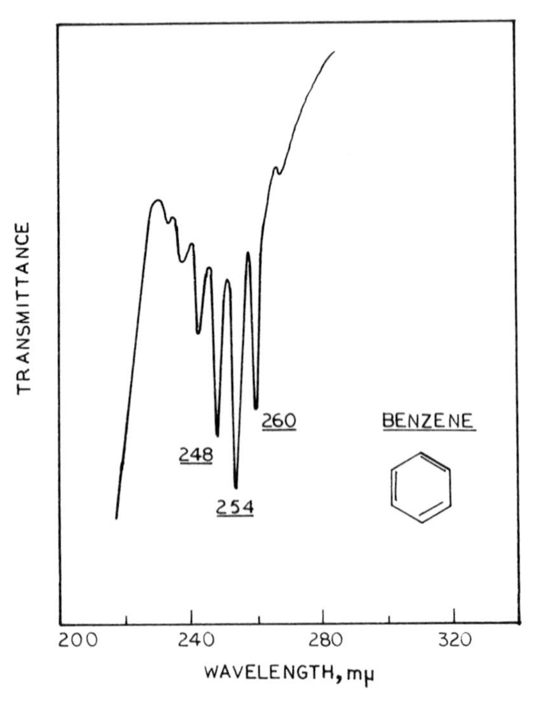

Figure 2. Ultraviolet absorption spectrum of benzene in ethanol

with available n-electrons or one with electron-withdrawing properties. This transition is assigned to the transfer of an electron to or from the benzene π orbital.

Longuet-Higgins and Murrell (*10*) describe such a transition as an electron transfer, symbolized as an E.T. absorption band. The $\pi \rightarrow \pi^*$ band of olefins and aromatic compounds and the $n \rightarrow \pi^*$ bands of carbonyl compounds are described as local excitation (L.E.) bands. Doub and Vanderbelt (*11*) studied the effect of various substituents on the absorption bands of benzene and attempted to relate the nature and pattern of substitution to the effect on the 203.5-μ band of benzene. By consideration of a number of disubstituted benzenes in hydroxylic solvents, those authors arrived at the empirical relationship (*12*) for prediction of the absorption bands of disubstituted benzenes.

$$\lambda_{max} = \Delta\lambda_o + 180 \quad (3)$$

The contributing substitutional parameters, $\Delta\lambda_o$, are found from Equation 4:

$$\delta\lambda_o' \times \delta\lambda_o'' = 24.05 \times \Delta\lambda_o \quad (4)$$

where $\delta\lambda_o$ is the displacing effect of each substituent.

This relationship holds reasonably well for p-disubstituted derivatives of benzene and is only applicable to substituents of opposite electronic effects, involving an o-p-directing group in the presence of an m-directing group. The combination of two similarly orienting groups results in a

Table II. $\delta\lambda_0$ Values for Benzene Substitution[a]

Substituent	$\delta\lambda_o$ (μ)	*Substituent*	$\delta\lambda_o$ (μ)
—H	24.05	—CN	43.9
—NH_3^+	25.6	—CO_2^-	44.7
—CH_3	29.6	—COOH	50
—Cl	29.3	—NH_2	50
—Br	31.5	—O^-	55
—OH	36.1	—$COCH_3$	65.3
—OCH_3	36.8	—CHO	68.9
—SO_2NH_2	37.5	—NO_2	91.6

[a] Taken from Daub and Vanderbelt, Ref. *11*.

spectrum similar to the stronger chromophore. Table II shows $\delta\lambda_o$ values for benzene substitution, and Table III shows the comparison between calculated and predicted values for some substituted benzenes of relevance to pesticide structures. In order to obtain the position of the second benzene absorption band (254 μ), the computed λ_{max} (displaced 203.5-μ band) is multiplied by a factor of 1.25 (*11*).

Table III. Prediction of the Position of λ_1 and λ_2 of Some Substituted Aromatic Compounds[a]

	λ_1		λ_2	
	Calculated	*Found*	*Calculated*	*Found*
Benzene	204	203.5	255	254
Toluene	209.6	206.5	262	261
Chlorobenzene	209.3	209.3	261.6	262
Nitrobenzene	271.6	268.5	–	–
Aniline	230	230	287.5	280
Phenol	216	210	270	270
Anisole	216.8	217	271	271

[a] Taken from Daub and Vanderbelt, Ref. *11*.

Table IV. Absorption Maxima of Some Chlorinated Hydrocarbon Pesticides and Related Compounds

Common Name	*Chemical Name*	λ_{max} (μ)	*Ref.*
Toluene		262	[a]
2–Phenyl ethanol		260	[a]
Phenyl acetic acid		258	[a]
Chlorobenzene		264	[a]
DFDT	1,1,1–Trichloro–2,2–bis (*p*-fluorophenyl)ethane	265	*13*
Perthane	2,2–Dichloro–1,1-bis(*p*-ethylphenyl)ethane	265	*14*
Kelthane	4,4′–Dichloro–α–(trichloromethyl)benzhydrol	265.5	*15*
Chlorobenzilate	Ethyl–4,4′–dichlorobenzilate	266	*13*
DDT	1,1,1–Trichloro–2,2–bis(*p*-chlorophenyl)ethane	267	*13*
DMC	4,4′–Dichloro–α–methylbenzhydrol	267.5	*13*
o,p–DDT		268.5	*13*
TDE	1,1–Dichloro–2,2–bis(*p*-chlorophenyl)ethane	268.5	*13*
o,o–DDT	1,1,1–Trichloro–2,2–bis(*o*-chlorophenyl)ethane	270	*13*
1,2,4–Trichlorobenzene		278	[a]
Fenac	Trichlorophenylacetic acid	276	[a]

[a] Present investigation.

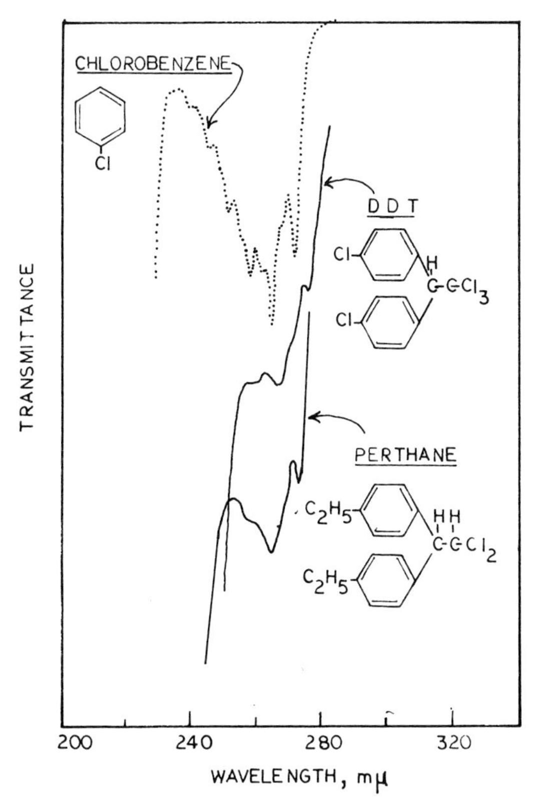

Figure 3. Ultraviolet absorption spectra of chlorobenzene, DDT, and benthane in ethanol

The pesticides that possess the typical benzoid chromophore are the chlorinated hydrocarbons. Alkyl or halogen substitution in the ring causes a bathochromic shift in the 254-μ band of benzene to the 260–270-μ region. Substitution in the side chain, however, does not cause any change to the spectrum in the near ultraviolet region. The spectra of some of these compounds are shown in Figure 3, and λ_{max} values for many representatives of this group are shown in Table IV. All these compounds are derivatives of toluene. Substitution in the side chain with an aromatic ring does not change the wavelength of maximum absorption—for example, diphenylmethane absorbs at 262 μ, as does toluene. However, substitution in the ring with halogen group causes a slight red shift (compare λ_{max} of toluene, DDT, DFDT, TDE, etc.). Increased halogen substitution causes more shift to longer wavelengths, as seen in phenylacetic acid, which absorbs at 258 μ while fenac (trichlorophenylacetic acid) absorbs at 276 μ.

Table V. Absorption Maxima of Phenol and Some Substituted Phenols

	λ_{max}	
	Neutral Medium	*Alkaline Medium*
Phenol	270	290
2,4–Dichlorophenol	284	302
2,4,5–Trichlorophenol	290	308
o–Cresol	271	290
4–Chloro–*o*–cresol	279	298
6–Chloro–*o*–cresol	272	294
4,6–Dichloro–*o*–cresol	281.5	305
m–Cresol	272	289
4–Chloro–*m*–cresol	279	299
6–Chloro–*m*–cresol	276	295
4,6–Dichloro–*m*–cresol	284	305
3,5–Xylenol	271	289
4–Chloro–3,5–xylenol	278.5	295
2,4–Dichloro–3,5–xylenol	287	302
2,4,6–Trichlor–3,5–xylenol	291	306

Table VI. Absorption Maxima for Some Phenolic Ether Pesticides

Common Name	*Chemical Name*	λ_{max} (μ)	*Ref.*
2–Phenoxyethanol		270	[a]
Methoxychlor	1,1,1–Trichloro–2,2–bis(*p*-methoxyphenyl)ethane	273	*13*
Aramite	2–(*p*–*tert*–butylphenoxy)–L-methylethyl–2–chloroethyl sulfite	275	*13*
Bis(*p*–chloro-phenoxy)methane		274	*13*
2,4–D	2,4–Dichlorophenoxyacetic acid	284	*12,18*
MCPP	2–(2–Methyl–4–chlorophenoxy-acetic acid)	287	*19*
MCPA	2–Methyl–4–chlorophenoxy-acetic acid	287	*19*
2,4,5–T	2,4,5–Trichlorophenoxy-acetic acid	289	*18*

[a] Present investigation.

PHENOLIC CHROMOPHORE. *Free Phenols.* Phenol exhibits a long-wavelength band at 270 μ (a_m 1450) which is considered to be the displaced and intensified local excitation band of benzene (254 μ). An intense absorption band appears at 210 μ (a_m 6000) which marks an E.T. transition (*4*) in which an *n*-electron of oxygen is transferred to the π-orbital of the ring. In an alkaline medium, an additional pair of nonbonded electrons is made available, and less energy is required for excitation, resulting in a shift of the wavelength of maximum absorption of about

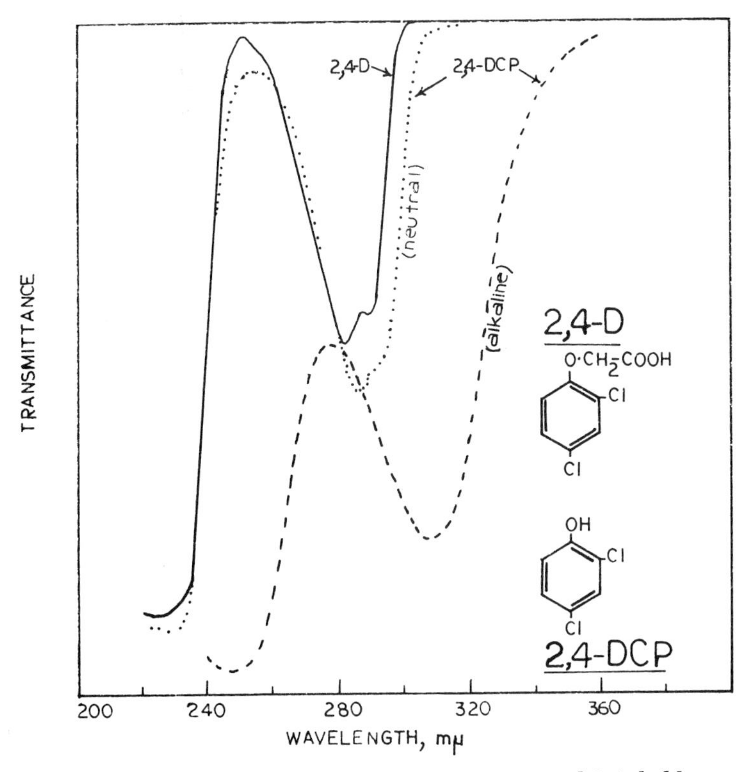

Figure 4. Ultraviolet absorption spectra of 2,4-D and 2,4-dichlorophenol (in neutral and alkaline medium) in ethanol

+20 μ. This bathochromic shift in alkaline media constitutes a useful method for the recognition of monohydric phenol chromophore and has been utilized in pesticide analysis (*16, 17*). Phenolic compounds constitute the degradation products of several pesticides, and several substituted phenols are used as fungicides and herbicides. Table V shows λ_{max} values of some phenolic compounds of importance in residue analysis.

Phenolic Ethers. Substitution of the phenolic proton by an alkyl group results in a spectrum which closely resembles that of the neutral species of the parent phenol (*4*). The pesticides that fall into this group are the phenoxy compounds. Table VI shows the λ_{max} values of some important pesticide ethers. The alkyl phenyl ethers do not exhibit a bathochromic shift with pH change into the alkaline range because there is no free phenolic proton to be lost and no charged anion is formed. This represents an important avenue in ultraviolet analysis, in which small amounts of free phenols may be determined in the presence of an excess of the corresponding phenolic ether. Figure 4 shows the spectrum of 2,4-D and its parent phenol, 2,4-dichlorophenol. In an alkaline medium,

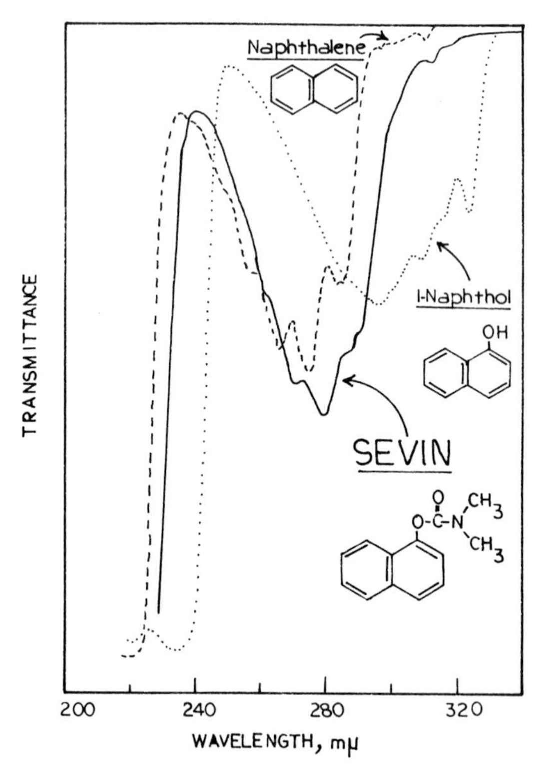

Figure 5. Ultraviolet absorption spectra of Sevin, 1-naphthol, and naphthalene in ethanol

only the phenol is shifted to longer wavelength, enabling its independent identification and determination.

Phenolic Esters. The esters of phenol do not absorb at the same wavelength as the parent phenol. The *n*-electrons of the oxygen atom are not available for transfer to the ring (as shown below) and a return to the spectrum of the parent hydrocarbon system is observed (*4*).

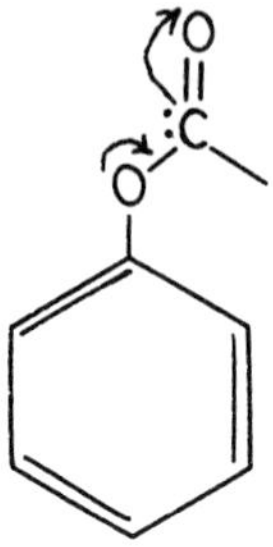

The pesticides that fall into this group are the *N*-alkyl carbamates and the phosphate esters. For example, the spectrum of Sevin (1-naphthyl-*N*,*N*-dimethyl carbamate) (Figure 5) reveals an absorption band at 280 μ which is only slightly shifted from that of naphthalene (λ_{max} 276). On the other hand, the absorption band of 1-naphthol at 296 μ is shifted by +20 μ to longer wavelength than that of naphthalene. Similarly, the phosphate ester pesticide paraoxon absorbs at 268 μ which is slightly shifted to longer wavelength than the absorption band of nitrobenzene (λ_{max} 260 μ) owing to the *p*-substitution in the ring. Substitution of P=O group by a P=S group in paraoxon results in a slight bathochromic shift, thus parathion absorbs at 274 μ. Table VII shows λ_{max} values of some pesticide esters together with their parent chromophores.

AROMATIC AMINE CHROMOPHORE. Aniline absorbs at 207, 230, and 280 μ, the central band having an intensity which is characteristic of an electron transfer (*4*). Generally, substitution in the ring or the hydrogen atoms of the amino group results in a shift of the 230-μ band of aniline to longer wavelengths. The anilides, the substituted phenylureas, and the *N*-phenylcarbamate pesticides are representatives of this group. Figure 6 shows the spectra of aniline and the substituted phenylureas, fenuron

Table VII. Absorption Maxima for Phenolic Ester Pesticides and their Parent Chromophores

Common Name	*Chemical Name*	λ_{max} (μ)	*Ref.*
Sevin	1–Naphthyl–*N*–methylcarbamate	280	*20*
Naphthalene[a]		276	*5*
Zectran	4–Dimethylamino–3,5–xylyl methylcarbamate	257	*21*
Dimethylaniline[a]		251	*4*
Baygon	*O*–Isopropoxyphenyl–*N*–methylcarbamate	270	*20*
Isopropoxybenzene[a]		270	[b]
Paraoxon	Diethyl–*p*–nitrophenyl phosphate	268	*22*
Parathion	*O*,*O*–diethyl–*O*,*p*–nitrophenyl phosphorothioate	274	*22*
Tri–*p*–nitrophenyl-phosphate		264	*22*
Ethyl di–*p*–nitro-phenylphosphate		267	*22*
Nitrobenzene[a]		260	[b]
Diazinon	*O*,*O*–diethyl–*O*–(2–isopropyl–4–methyl–6–pyrimidinyl) phosphorothioate	247.5	*23*
Pyrimidine[a]		243	*4*

[a] Parent chromophore of the preceding pesticide(s).
[b] Present investigation.

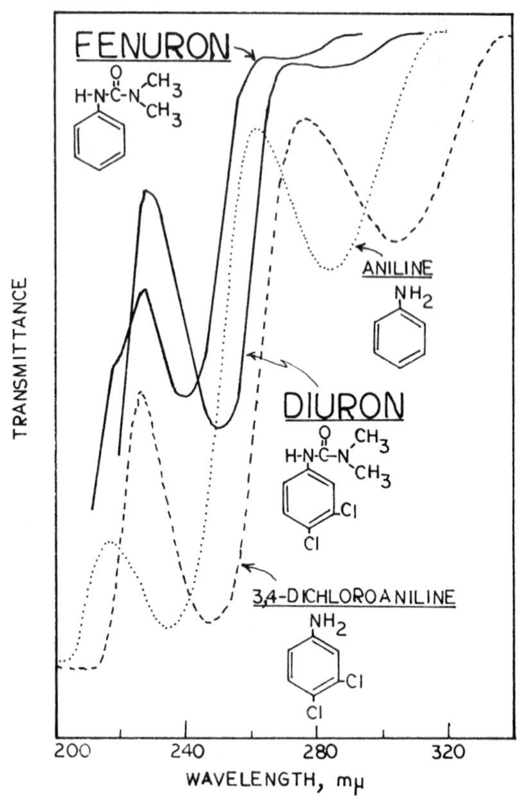

Figure 6. Ultraviolet absorption spectra of aniline, 3,4-dichloroaniline, fenuron, and diuron in ethanol

Table VIII. Absorption Maxima of Some Aniline Derivative Pesticides and Related Compounds

Common Name	*Chemical Name*	λ_{max}	*Ref.*
Aniline		238	[a]
IPC	Isopropyl–*N*–phenylcarbamate	234	*19*
CIPC	Isopropyl–*N*–(3–chlorophenyl)-carbamate	237.5	*19*
Fenuron	1,1–Dimethyl–3–phenylurea	237	[a]
3,4–Dichloroaniline		247	[a]
Propanil	3′,4′–Dichloropropionanilide	248	*19*
Monuron	3–(*p*–Chlorophenyl)–1,1–dimethyl-urea	244	[a]
Diuron	3–(3,4–Dichlorophenyl)–1,1–dimethylurea	246	[a]
Neburon	3–(3,4–Dichlorophenyl)–1–butyl–1–methylurea	245	[a]

[a] Present investigation.

and diuron. The wavelengths of maximum absorbance of several other representatives of aniline derivatives are shown in Table VIII.

Heterocyclic Compounds. PYRIDINE CHROMOPHORE. The spectrum of pyridine in ethanol reveals an absorption band at 247 μ which is considered the counterpart of the 254 μ ($\pi \rightarrow \pi^*$) band of benzene. In aqueous solutions, pyridine absorbs at 251 μ and a new band at 270 μ appears which marks an $n \rightarrow \pi^*$ transition of an n-electron to the ring orbital (*4*). Acidification of a solution of compound containing the pyridine nucleus removes this long-wavelength band from the spectrum. Substitution of the pyridine ring in the α or β positions results in a shift of the 257-μ band to longer wavelengths. Nicotine (Figure 7) and anabasine are typical examples of the β-substituted pyridyl chromophore which absorbs at 262 μ. Paraquat (aqueous) absorbs at 256 μ. In this molecule there are no nonbonded electrons but they are involved in the cation formation, and the spectrum is very similar to that of an acidic solution of

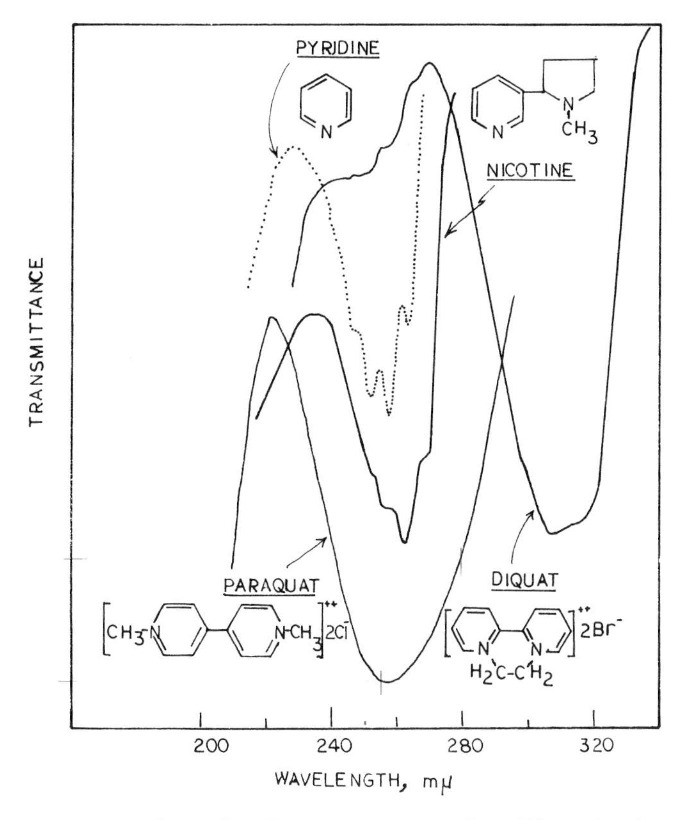

Figure 7. Ultraviolet absorption spectra of pyridine, nicotine (in ethanol), paraquat, and diquat (in water)

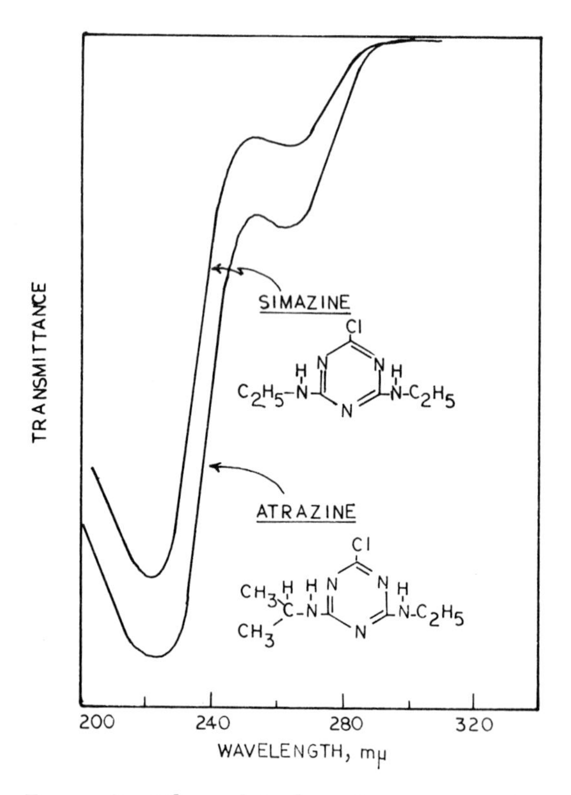

Figure 8. Ultraviolet absorption spectra of simazine and atrazine (in water)

pyridine. The spectrum of diquat, on the other hand, shows a dramatic bathochrome shift to 308 μ similar to that observed with the α-*ter*-amino-pyridyl derivatives (*24*).

Sym-TRIAZINE CHROMOPHORE. The spectrum of simazine and atrazine in water (Figure 8) shows a weak band at 263 μ which is to be compared with the $\pi \rightarrow \pi^*$ (255 μ) band of benzene. Another intense band at 220 μ is observed which is characteristic of all the substituted *sym*-triazines (Table IX).

Ultraviolet Micro Spectrophotometry as an Aid to the Identification of Pesticide Residues

Recently Suffet (*25*) and Faust and Suffet (*26*) reported an intensive study on the separation and identification of the phosphate ester pesticides parathion, diazinon, and fenthion and their degradation products. A summary of a microspectrophotometric ultraviolet technique that was utilized as an aid in identifying these compounds is presented here.

A Beckman DK-2A spectrophotometer was calibrated with a standard benzene vapor spectrum. The spectrophotometer was adjusted for the highest resolution with a tolerable signal-to-noise ratio. Since high resolution demands a narrow slit width, instrument sensitivity was set as high as possible. As a maximum signal response was needed, a low time constant was set, and a relatively slow scanning speed was used.

A 1-cm Beckman microcell (aperture volume of 50 μl) of fused silica windows with a range of 220 to 2500 μ was utilized. A variable-beam condenser served to attenuate the reference beam. An actual wavelength cutoff at 230 μ was observed for the microcell. In order to lower the cutoff point, a matched 1-cm standard silica cell filled with solvent was centered behind the reference beam attenuator. This allowed a lower cutoff at 205 μ. Ethanol, 95%, was chosen because of a low cutoff point, 205 μ, and a high dissoluble capacity for the compounds of interest.

The samples were collected from GLC columns by two techniques. A system with KBr as an absorbent was used to collect GLC peaks from single-column operation with a flame ionization detector. This was adapted from an original design for dual column operation (*27*). It consisted of column effluent splitter, collection tube, and fraction collector. The splitting ratio of the flow was 1:1.7, detector:output. This system was chosen for the following reasons:

(*a*) There is a direct GLC to microspectrophotometric operation with minimum intermediate handling. This eliminates contamination problems in handling microgram quantities.

(*b*) The collection of several peaks during one chromatographic run is possible.

(*c*) The ease of operation and repeatability.

(*d*) Collected fractions can be dissolved and rechromatographed.

(*e*) Recoveries of 50 to 75% have been obtained.

The GLC peak collected on the KBr may be inserted into the microcell by either of two techniques: by pouring the powder directly into a

Table IX. Absorption Maxima for Some *sym*-Triazine Pesticides[a]

Common Name	*Chemical Name*	λ_{max} (μ)
Atratone	2–Ethylamino–4–isopropylamino–6–methoxy–*s*–triazine	220
Simeton	2,4–Bis(ethylamine)–6–methoxy–*s*–triazine	220
Prometon	2,4–Bis(isopropylamino–6–methoxy)–*s*–triazine	220
Simazine	2–Chloro–4,6–bis(ethylamino)–*s*–triazine	222
Propazine	2–Chloro–4,6(isopropylamino)–*s*–triazine	223
Trietazine	2–Chloro–4–diethylamino–6–ethylamino–*s*–triazine	228
Atrazine	2–Chloro–4–ethylamino–6–isopropylamino–*s*–triazine	222

[a] Taken from Baily and White, Ref. *19*.

Table X. Comparison of the Ultraviolet Wavelength

Compound	*Standard 1-cm Matched Cells, 95% EtOH, μ*	*50-μl Microcell, 95% EtOH, μ*
Diazinon	247.5	247.5
Diazoxon	245	246
IMHP	271.5, 225	272
Parathion	274	275
Paraoxon	271	270.5
p–Nitrophenol	313, 231	314
Baytex	sh, 252	252
Bayoxon	sh, 248.5	249

solvent in the microcell or by dispersing an infrared micropellet directly into a solvent within the microcell. The wavelength cutoff point was higher for ethanol and KBr (220 μ) than for ethanol alone (205 μ). The solvent spectra were recorded before the spectra of the standard, and collected samples were run in order to assure cell cleanliness.

Table X shows a comparison between the ultraviolet spectral maxima obtained in this study and the literature values. The standard and collected GLC spectra are qualitatively the same. The characteristic shift to lower wavelength by an organophosphate oxon is evident. The shift to a lower wavelength when a phosphate ester replaces the H atom of the hydrolysis product is noted also (Diazinon).

These compounds can be identified directly from an electron capture detector outlet, as the general sensitivity of the detector is in the nanogram-to-microgram range. Figure 9 shows an ultraviolet spectrum of IMHP which was collected from an electron capture detector. The collection device of Kabot (*31*) was used with an aged liquid phase and an absorbent in a melting point tube as suggested by Amy *et al.* (*32*). The compound was eluted from the liquid phase directly into the microcell with 95% ethanol. Cleanliness of the collection device was assured by eluting the glass wool retainer and melting point tube with ethanol and drying it before collecting.

These collection and microspectrophotometric techniques can be applied to many other pesticides that can be separated by gas–liquid chromatography.

If 95% transmittance (0.05 absorbance) above the noise level is assumed as a minimum acceptable response and if the values of l, a_m, c_m, and molecular weight (MW) are known, then the theoretical lower limit of detectability (LLD) in ng may be calculated for a microcell with a volume of 60 μl.

Maxima of the Diazinon, Parathion, and Baytex Systems[a]

Collected on KBr from GLC Microcell, μ	*Literature Value, μ*	*Ref.*
247.5	247.5	*23*
245.5		
272	272, 224.5	*23*
274	274 (276)	*28*
270	270 (268)	*29*
314	314	*28*
252	252	*30*
249		

[a] All maxima are corrected for wavelength calibrations.

$$A = a_m c_m l \quad (5)$$

$$0.05 = a_m \times \frac{ng\ 1000}{60 \times MW} \times 1.0$$

$$\mathrm{LLD} = ng = \frac{MW}{a_m} \times 3 \times 10^3$$

Table XI shows these theoretical calculations for the organophosphates of interest.

Whereas microgram quantities are needed for infrared examination, only nanograms are needed with the ultraviolet technique (Table XI). Hence, any compound which is collected in sufficient amounts for infrared examination will be sufficient for ultraviolet examination.

Recently, Kirkland (*33, 34*) described a high-performance ultraviolet photometric detector for use with efficient liquid chromatographic columns. The key advantages of this device are its unusually high sensitivity, stability, and wide range of linearity. The detector incorporates a low-volume sample cell which permits its use with efficient, small-bore analytical columns. Chromatographic peaks containing nanogram amounts of moderate-absorptivity compounds could be observed readily with the detector, and its range of linearity was roughly comparable with thermal conductivity detectors for gas chromatography. Total volume of the micro flow-through cells, including a 10-cm length of 0.02-inch inlet tubing, were 20 and 7.5 μl, respectively. The detector was capable of operating at a full-scale sensitivity of 0.01 absorbance unit (at 254 μ) with a short-term noise of about ±0.00015 absorbance. This UV photometric detector was capable of detecting extremely small amounts of materials which absorb at 254 μ even though these components may have their maximum absorption at much higher or lower wavelengths. While

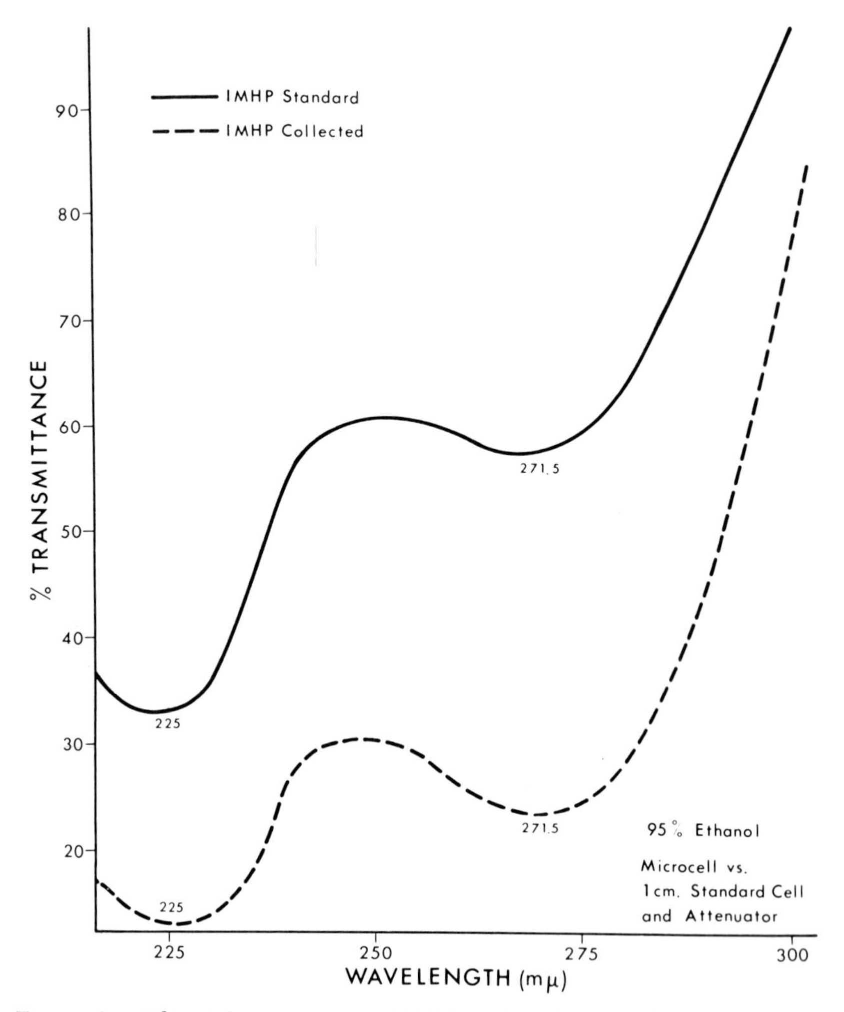

Figure 9. Ultraviolet spectra of IMPH collected from electron capture detector outlet

operation at 254 μ has proved to be the most generally useful, the detector could be operated at other wavelengths by changing to appropriate sources and filters.

Ten nanograms of diuron (absorptivity = 88 liter/gram-cm at 254 μ) were readily detected (Figure 10) with the detector operating at a full-scale sensitivity of 0.01 absorbance. The large peak preceding diuron is a solvent impurity in the 100-μl sample aliquot injected into the chromatographic column. In Figure 11, a 3-μl aliquot containing 1 μg of fenuron,

Table XI. Calculated Lower Limit of Ultraviolet Detectability (ng) for the Diazinon, Parathion, and Baytex Systems

Compound	*MW*	a_m	*Calculated LLD, ng*	*Experimental LLD, ng*[a]
Diazinon	304	4,300	210	–
IMHP	153	4,800	96	180
Parathion	291	9,570	90	–
Paraoxon	275	9,630	–	–
		9,590	84	–
p–Nitrophenol	139	10,750	39	66
Baytex	278	12,100	69	–
Bayoxon	262	13,500	58	–
MMTP	154	–	–	–

[a] Determined in 60–μl microcell.

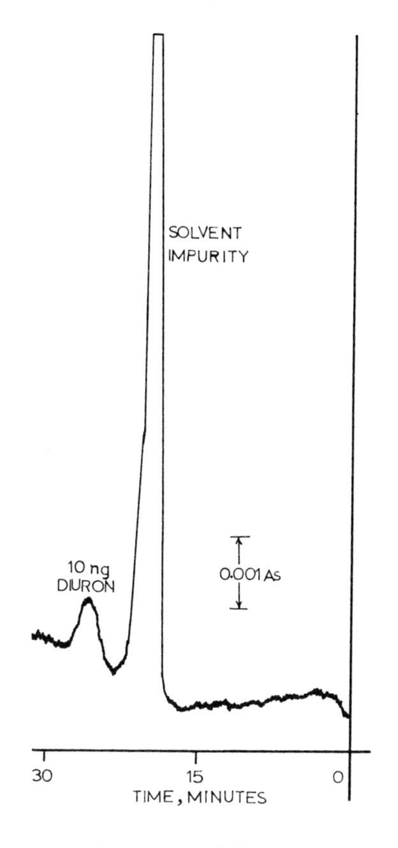

F. A. Gunther and R. C. Blinn, "Analysis of Insecticides and Acaricides," Interscience

Figure 10. High-sensitivity detection

Sample: 100 μl of 0.1 μg/ml diuron [3-(3,4-dichlorophenyl)-1,1-dimethylurea] in n*-butyl ether; sensitivity: 0.01; absorbance: full scale; carrier:* n*-butyl ether; flow rate: 0.26 cc/min (after Ref.* 13*)*

monuron, diuron, and linuron was chromatographed using a detector sensitivity 1/20 of maximum.

Although obviously limited to compounds having ultraviolet absorption, the UV detector described appears particularly applicable for quan-

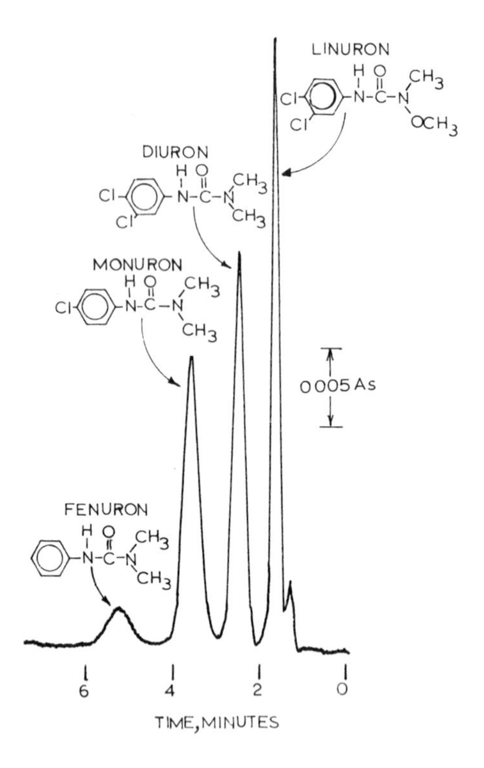

Journal of Chromatographic Science

Figure 11. Separation of substituted urea herbicides

Column: 500 mm × 2.1 mm, i.d.; packing: 1.0% β,β′-oxydipropionitrile on 37–44 micron CSP support; carrier: dibutyl ether; flow rate: 1.14 cc/min; sample: 1 μl of 67 μg/ml each in dibutyl ether (after Ref. 34)

titative analysis; still, it is necessary to calibrate individually for each of the compounds to be analyzed. The very high sensitivity of the device makes it particularly useful for trace analysis. Satisfactory use of the UV photometric detector requires that the chromatography be carried out with spectrally pure carrier solvents; however, a wide range of such materials is now readily available from commercial sources.

Summary

Ultraviolet spectrophotometry can be considered a valuable tool for confirming tentative identification of pesticide residues. Because of its extreme sensitivity, submicrogram quantities are often sufficient to obtain reasonable spectrograms. A knowledge of the correlation between the UV spectrum and structure may provide clues about the general type of chromophore present. The transparency of many groupings (and often large segments of complex molecules) in the ultraviolet region imposes a limitation of usefulness on the results of interpretation of absorption bands in this region. However, when taken in conjunction with the wealth of detail provided by infrared, NMR, and mass spectra, UV spectrophotometry may lead to structural proposals of value to the pesticide analyst.

The high absorptivity values of the UV-absorbing compounds, particularly at the lower wavelengths (200–220 μ), makes ultraviolet spectrophotometry a powerful analytical technique suitable for determination of pesticide residues in the submicrogram level. The development of high-performance ultraviolet photometric detectors in connection with low-volume cells and long light path provides a versatile and convenient means of analyzing low concentrations of pesticide residues with sensitivities that can be matched with many GLC detectors.

Literature Cited

(1) Blinn, R. C., Gunther, F. A., *Residue Rev.* (1963) **2,** 99.
(2) Chakraborty, B. B., Long, R., *Environ. Sci. Technol.* (1967) **1,** 828.
(3) Crummett, W., *Anal. Chem.* (1966) **38,** 410R.
(4) Scott, A. I., "Interpretation of the Ultraviolet Spectra of Natural Products," Macmillan, New York, 1964.
(5) Friedel, R. A., Orchin, M., "Ultraviolet Spectra of Aromatic Compounds," Wiley, New York, 1951.
(6) Barrow, G. M., "The Structure of Molecules," p. 21, Benjamin, New York, 1964.
(7) Conover, L. H., *Chem. Soc. Spec. Publ.* (1956) **5,** 48.
(8) Ungnade, H. E., *J. Am. Chem. Soc.* (1953) **75,** 432.
(9) Woodward, R. B., *J. Org. Chem.* (1941) **63,** 1123.
(10) Longuet-Higgins, H. C., Murrell, J. N., *Proc. Phys. Soc.* (1955) **68A,** 60.
(11) Doub, L., Vanderbelt, J. M., *J. Am. Chem. Soc.* (1947) **69,** 2714.
(12) Aly, O. M., Faust, S. D., *J. Am. Water Works Assoc.* (1963) **55,** 639.
(13) Gunther, F. A., Blinn, R. C., "Analysis of Insecticides and Acaricides," Interscience, New York, 1955.
(14) Weiss, A. R., "Determination of Parthane Residues," Rohm & Haas Co., mimeo, 1955.
(15) Gunther, F. A., Blinn, R. C., *J. Agr. Food Chem.* (1957) **5,** 517.
(16) Aly, O. M., El-Dib, A. M., *Water Res.*, in press.
(17) Ketelaar, J. A. A., *Rec. Trav. Chim.* (1950) **69,** 649.
(18) Gordon, N., Beroza, M., *Anal. Chem.* (1952) **24,** 1968.
(19) Bailey, G. W., White, J. L., *Residue Rev.* (1965) **10,** 97.
(20) Aly, O. M., El-Dib, A. M., *Proc. Fifth Rudolfs Res. Conf.*, in press, 1969.
(21) Eberle, D. O., Gunther, F. A., *J. Assoc. Offic. Agr. Chemists* (1965) **48,** 927.
(22) Gersmann, H. R., Ketelaar, J. A. A., *Rec. Trav. Chim.* (1958) **77,** 1018.
(23) Blinn, R. C., *J. Agr. Food Chem.* (1955) **3,** 1013.
(24) Furman, W. B., *J. Assoc. Offic. Anal. Chemists* (1968) **51,** 1111.
(25) Suffet, I. H., Ph.D. thesis, Rutgers University, New Brunswick, N. J., 1968.

(26) Faust, S. D., Suffet, I. H., "Microorganic Matter in Water," ASTM-STP 448, p. 24, American Society for Testing Materials, 1969.
(27) Guiffrida, L., *J. Assoc. Offic. Agr. Chemists* (1965) **48,** 354.
(28) Biggs, A. I., *Analyst* (1955) **80,** 279.
(29) Williams, E. F., *Ind. Eng. Chem.* (1951) **43,** 950.
(30) Ibrahim, F. B., Cavegnol, J. C., *J. Agr. Food Chem.* (1966) **14,** 369.
(31) Kabot, F., *G.C. Newsletter,* Perkin Elmer Corp. (1967) **3,** 1.
(32) Amy, J. W., Chait, E. M., Baitinger, W. E., McLafferty, F. W., *Anal. Chem.* (1965) **37,** 1265.
(33) Kirkland, J. J., *Anal. Chem.* (1968) **40,** 391.
(34) Kirkland, J. J., *J. Chromatog. Sci.* (1969) **7,** 7.

RECEIVED June 12, 1970. This is a paper of the Journal Series, New Jersey Agricultural Experiment Station, Rutgers, The State University of New Jersey, Department of Environmental Sciences, New Brunswick, N. J. 08903.

8

Past, Present, and Future Application of Paper and Thin-Layer Chromatography for Determining Pesticide Residues

MELVIN E. GETZ

Entomology Research Division, Agricultural Research Service, U. S. Department of Agriculture, Beltsville, Md. 20705

The past and present uses of paper and thin-layer chromatography have been confined largely to qualitative identification and semiquantitative estimation. Developments in optical scanning instrumentation and refinements in spotting techniques have endeavored to show that these methods of analysis can be quantitative. The precision and accuracy of this approach are dependent upon uniformity of chromatographic layer, uniform spotting of samples and standards, and uniform application of chromogenic reagents. These new investigations were undertaken so an alternate quantitative method of analysis would be available for pesticide residue analysis.

The effect on the environment of pesticides and their residues has become a growing concern of the scientific community and the general public (*1*). Pesticides embrace a large variety of chemicals, including insecticides, fungicides, herbicides, rodenticides, nematocides, and molluscicides. Since these compounds are toxic in nature, many countries regulate their use by law and specify how much residue can be left in a food product. In order to enforce such regulations, sophisticated analytical methods have had to be developed.

In addition to food analysis, the environment is monitored to measure the degree of contamination that might be caused by the use of pesticides. These projects entail sampling of wild life, water sources, and soils. Wealthy countries can utilize highly sophisticated and expensive equipment, such as gas chromatography coupled with mass spectrometry, for residue determination and identification. However, since there is international concern about the use of pesticides, there is a need for a reliable analytical method that is simple and inexpensive.

Paper and thin-layer chromatography are two such approaches. The techniques are simple and require inexpensive equipment. If one chooses the proper solvents, adsorbents, and chromogenic reagents, residues can be isolated on a sheet of chromatographic paper or a thin layer of adsorbent. The type of residue can then be determined by comparison with reference standards.

The applications of these techniques are reviewed in this paper. The author's personal experiences are presented with a look into future applications, in particular, quantitative determinations by optical scanning methods.

Review

Paper Chromatography. The first important step for any pesticide residue methodology is to extract the residue from the substrate and isolate it in a pure enough state so that it can be identified and measured. This purification step is called "clean-up," and there is a great quantity of literature concerning the different approaches (*2, 3, 4, 5, 6*).

Once the residue has been isolated in a fairly pure state, it can be further resolved and identified by chromatography. Paper chromatography, which was introduced by Consden *et al.* (*7*) in 1944, was the first technique used. Initially, it was chiefly an art, since it depended on manipulations of the individual investigators. By 1957, suppliers of chromatographic papers started to manufacture a high-quality product. It was at this time that Mitchell (*8*) developed paper chromatography into a science by setting up a systematic approach for determining optimum conditions for resolution and sensitivity. His work with immobile–mobile phase chromatography (*9*) demonstrated that pesticide separations could be obtained with a wide degree of resolution and flexibility.

One of the first laboratory groups to apply paper chromatography as an identification and estimation technique was that of Müller *et al.* (*10*). In 1957, at the Control Laboratory in Basel, his group developed an extraction and clean-up procedure for a few organochlorine and organophosphorus insecticides and then determined the quantity of residue with reverse phase paper chromatography. Similar procedures were followed by McKinley and Mahon in Canada (*11*) and Mills in the United States (*12*). This first series of papers triggered wide applications of paper methods for residue determinations (*13, 14, 15, 16, 17, 18, 19, 20, 21, 22*).

Thin-Layer Chromatography. A few years after the introduction of paper chromatography, the principles of thin-layer chromatography were demonstrated by Meinhard and Hall (*23*). Kirchener *et al.* (*24*) introduced the initial thin-layer coating techniques. Sola *et al.* (*25*) were the first group to show that insecticides could be chromatographed by this technique, with Walker and Beroza (*26*) presenting data obtained by a comprehensive coverage of a large number of solvents with many types of insecticides. The successful application of thin-layer chromatography to determine multiple organochlorine and organophosphorus residues in food products was accomplished by Kovacs (*27, 28*).

Excellent reviews of TLC techniques as applied to pesticides have been written by Conkin (*29*) and by Abbot and Thomson (*30*).

Personal Techniques

Clean-up. Charcoal is the favorite adsorbent for the first steps in the author's clean-up procedures (*31, 32*). It is an excellent decolorizing agent and removes much of the large-molecule interferences. When it was initially used, there was a problem of reproducibility. Heavy metal cations, especially iron, were chelating with or tying up the insecticides

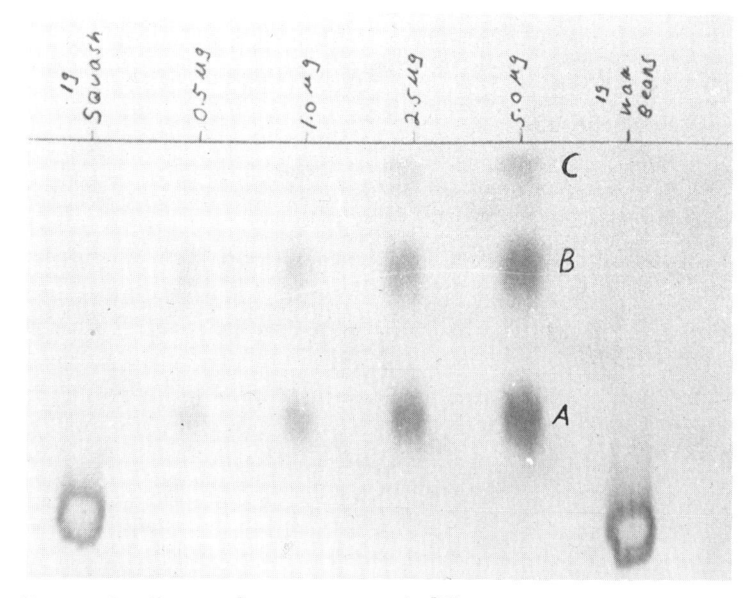

Figure 1. Paper chromatogram of different concentrations of a mixture of methyl parathion (A), Methyl Trithion (B), and carbophenothion (C); also, 1-gram aliquots of cleaned-up samples of control squash and wax beans

Paper: Whatman #1 20- × 20-cm
Immobile phase: 20% dimethylformamide
Mobile phase: 2,2,4-trimethylpentane
Chromogenic reagent: p-nitrobenzylpyridine

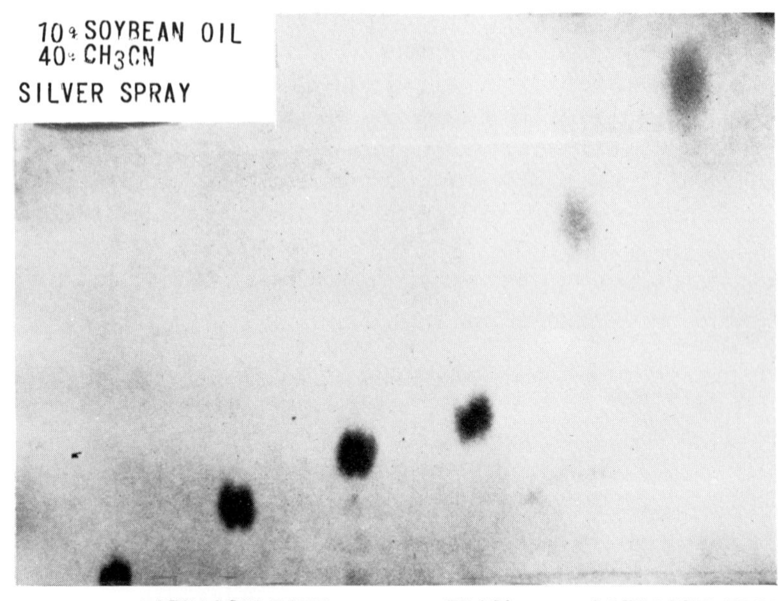

Figure 2. Paper chromatogram of 10 μg of carbophenothion (Trithion) and its oxidative metabolites

Paper: Whatman #3MM 20- × 20-cm
Immobile phase: 10% soybean oil
Mobile phase: 40% acetonitrile in distilled water
Chromogenic reagent: silver nitrate–bromcresol green

so that none or only part of them were being eluted. A simple treatment with hydrochloric acid tied up the metal ions and reproducible results were obtained then from the different wood charcoals. Both column elution and batchwise clean-up techniques are used.

If there is a significant amount of oily material left after the decolorizing step, a second treatment is necessary, involving a small-volume liquid–liquid partition (*33*). This approach has worked well for organochlorine and organophosphorus insecticides.

Spotting. The present investigations of this laboratory are directed toward making thin-layer and paper techniques quantitative by the use of optical scanning. In particular, applications of the reflectance scanning device of Beroza *et al.* (*34*) are being explored. Variations in the application of samples and standards to the chromatographic media have a pronounced effect on the linearity of standard curves and do not give good reproducibility.

For qualitative work and the first quantitative scanning techniques, the sample was evaporated to dryness in a centrifuge tube and then transferred to the chromatogram with a spotting pipette. Several rinsings were performed with a volatile solvent such as ether or chloroform. The reference standard was usually spotted with a few μl of solvent. The final result was a diffuse spot for the sample and a tight spot for the standard. After development of the chromatogram and visualization of the spots, it was difficult to make an accurate comparison with the standard.

This problem is minimized by the use of an automatic spotter designed by Getz (*35*). This spotting device, allowing complete control of the initial spot size of both sample and standard by the appropriate choice of solvents and air flow, is an improvement over an earlier design by Beroza *et al.* (*6*).

Chromatographic Media and Solvent Systems. Whatman #1 and #3MM 20- × 20-cm papers are used for paper chromatography. The reverse phases of Mitchell (*9*) give excellent results for the organochlorine and parent organophosphorus insecticides. The reverse phases of Getz (*36*) successfully resolve the oxidative metabolites of the organo-

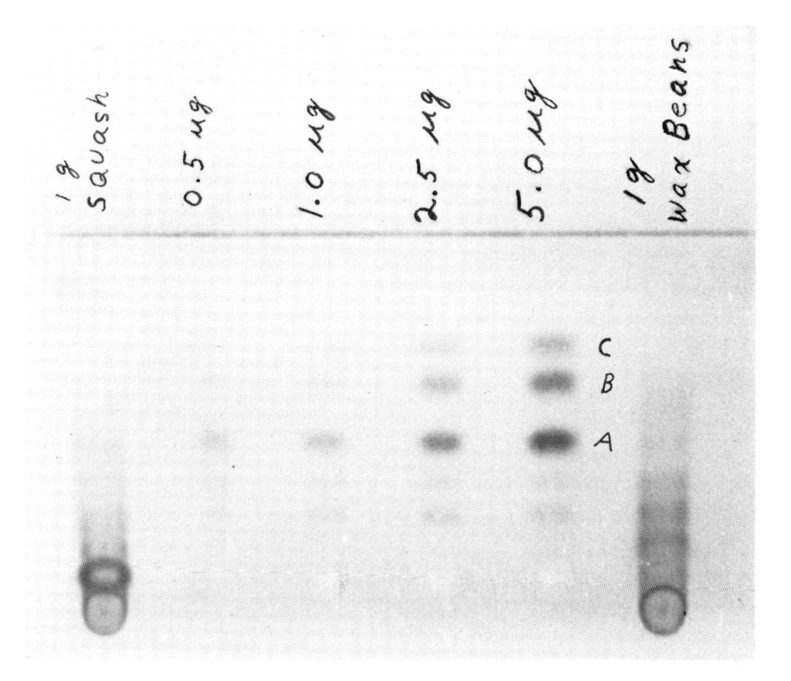

Figure 3. Thin-layer chromatogram of different concentrations of a mixture of methyl parathion (A), Methyl Trithion (B), and carbophenothion (C); also, 1-gram aliquots of cleaned-up samples of control squash and wax beans

Plate: Brinkman prepared silica gel, 250 mμ
Solvent: 70% 2,2,4-trimethylpentane, 25% acetone, 5% chloroform
Chromogenic reagent: p-*nitrobenzylpyridine*

Figure 4. Thin-layer chromatogram of actual organochlorine residues in nursery soils

Plate: Eastman silica gel sheet
Solvent: 90% 2,2,4-trimethylpentane, 10% methylene chloride
Chromogenic reagent: silver nitrate–phenoxyethanol

phosphorus insecticides. Figure 1 represents some organophosphorus insecticides chromatographed with a nonaqueous mobile phase, and Figure 2 represents the oxidative metabolites of carbophenothion chromatographed by an aqueous system.

Silica gel, alumina, Florisil, cellulose, and porous glass are used for preparing 20- × 20-cm glass plates. Calcium sulfate, colloidal silica, and colloidal alumina are used as binders. Commercially-prepared plates also have been used. The solvent systems and adsorbents investigated by Getz and Wheeler (*37*) are used for identifying organophosphorus insecticides. The organochlorine insecticides are resolved by alumina and silica gel, using the solvent systems of Kovacs (*27*), Walker and Beroza (*26*), and Beroza *et al.* (*38*).

Figure 3 shows separation of some organophosphorus insecticides by thin-layer chromatography, and Figure 4 shows a chromatogram of some organochlorine insecticides in soil extracts.

Development is carried out in tanks without liners, for experience has shown that the shape of the spot is more symmetrical if there is no liner present, especially when binary or ternary solvent systems are used.

Chromogenic Reagents. In order to determine the R_f values of pure insecticides, any type of reagent may be used to visualize the migrated spot. But when this reagent is used for actual residue determinations, there may be nonresidue spots formed by extractives from the substrate. Therefore, it is desirable that the chromogenic reagents be as selective as possible.

For detecting the presence of organochlorine compounds, an acetone solution of silver nitrate is used as a spray reagent (*8*). Silver nitrate also can be incorporated into the thin-layer material (*39*) and seems to give better results. However, the thin-layer media must have a low chlorine content or else the background may turn brown or gray. The spots

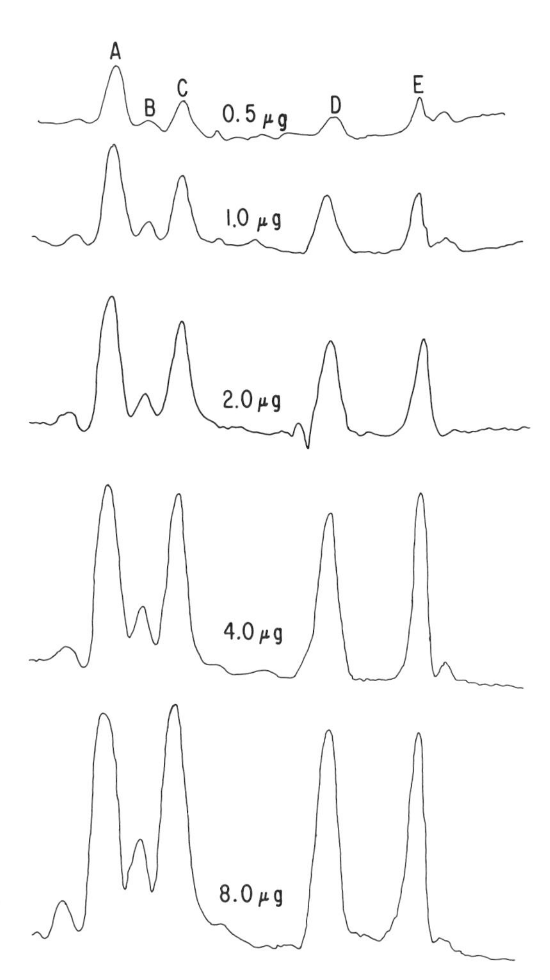

Figure 5. Scan of five different concentrations of an organophosphorus insecticide mixture: (A) dimethoate, (B) mevinphos, (C) Dasanit, (D) malathion, (E) carbophenothion

Plate: Quanta prepared silica gel, high abrasion resistance
Solvent: 70% 2,2,4-trimethylpentane, 25% acetone, 5% chloroform
Chromogenic reagent: p-*nitrobenzylpyridine*

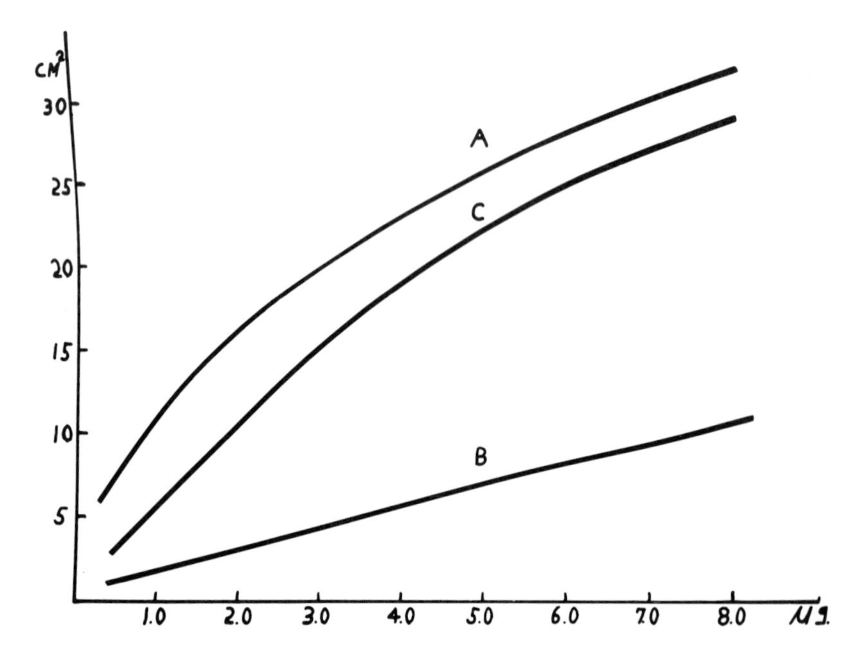

Figure 6. Standard curves obtained from (A) dimethoate, (B) mevinphos, (C) Dasanit in the concentration range of 0.5 to 8.0 μg

are developed by exposing the chromatogram to high-intensity UV light. The colors vary from brown to purplish-black.

Several reagents are used for the organophosphorus insecticides. An acetone solution of silver nitrate and bromcresol green (or bromphenol blue) (*31, 32*) is sprayed onto the chromatogram. The spots are visualized by spraying with a 0.01% citric acid solution or a pH 4.0 citrate buffer solution. This reagent reacts only with the thiophosphoryl configurations to give blue or magenta spots.

Treatment with *p*-nitrobenzylpyridine and tetraethylenepentamine visualizes all the organophosphorus insecticides as blue or magenta spots (*40*).

Serum cholinesterase with bromthymol blue indicator and acetylcholine as a substrate detects the cholinesterase-inhibiting insecticides (*36*).

Possible interferences with the silver nitrate chromogenic reagent are sulfur compounds which would give a brown spot and reducing groups on molecules which might produce a spot.

To date, no compounds other than organothiophosphates give a blue or magenta spot with the silver nitrate–dye reagent. Sulfur compounds will react with the silver nitrate to give brown spots, and artifacts with acid characteristics having R_f's similar to the insecticides would prevent the color formation.

The *p*-nitrobenzylpyridine will react with epoxides, lactones, other alkylating-type compounds, and herbicides such as atrazine. It will also give a color reaction with organochlorine insecticides that have an epoxy group.

Quantitative Chromatography

There have been many attempts at quantitating paper and thin-layer chromatograms (*41*), but the simplest approach appears to be optical density measurements of the migrated spots. Many excellent scanning devices are now on the market. They are precise optical instruments capable of great accuracy where the optical measurements are concerned. The one used in this laboratory is the reflectance scanner of Beroza *et al.* (*34*) which uses double-beam fiber optics.

However, there are other manipulations performed for chromatography that affect the precision and accuracy of the final result. As mentioned previously, the spotting technique variation was minimized by the use of an automatic spotter (*35*) which spots sample and standard under similar parameters, and the size of the initial spots are adjusted so that the Einstein-Smoluchowsky effect is minimized (*42*), allowing a range of concentrations from the minimum detectable to 20 μg.

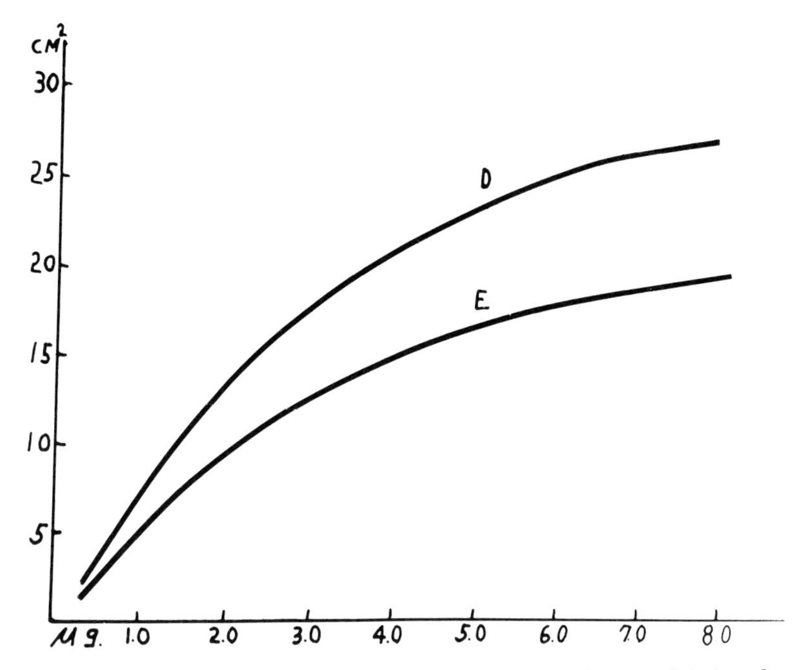

Figure 7. Standard curves obtained from (D) malathion and (E) carbophenothion in the concentration range of 0.5 to 8.0 μg

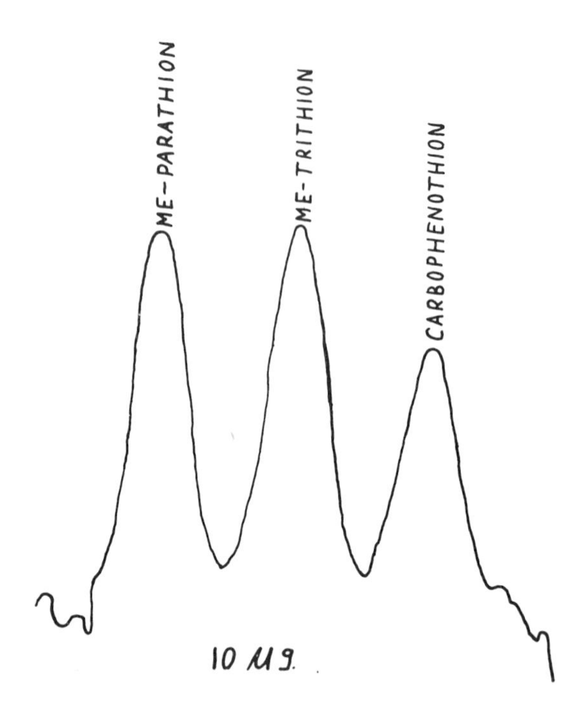

Figure 8. A reflectance scan of 10-μg mixture of three organophosphorus insecticides resolved with paper chromatography

Paper: Whatman #1 20- × 20-cm
Immobile phase: 20% dimethylformamide
Mobile phase: 2,2,4-trimethylpentane
Chromogenic reagent: p-*nitrobenzylpyridine*

With the spotting technique optimized, the reproducibility precision was almost entirely dependent upon the uniformity of the chromatographic layer. Most plates had thick edges so that samples or standards spotted near the edge could not be included in the quantitative calculations. One brand of commercial plate prepared by spraying techniques gave the best uniformity from edge to edge, and when standard curves were prepared using the whole plate, very smooth curves were obtained. The other commercial plates used gave good results when the center was used for four samples or standards.

The abrasive resistance of the plates have to be better than those used for qualitative work, and the chromogenic reagents have to be applied by dipping or incorporating into the thin layer to prevent damage to the surface and uneven spotting of the background area.

Figure 5 shows the scans obtained from different concentrations of a mixture of five organophosphorus compounds: dimethoate (A), mevinphos (B), Dasanit (C), malathion (D), and carbophenothion (E). Figures 6 and 7 show the standard curves obtained from these scans by plotting the areas of the curves *vs.* concentration.

Figure 8 is a scan of the 10-μg mixture of methyl parathion, Methyl Trithion, and carbophenothion shown in Figure 1.

Figure 9 compares the standard curves obtained from various concentrations of diazinon at two different attenuation settings. The attenuator response is linear but the reflectance scan does not respond linearly (5).

Future Prospects

As we compare paper with thin-layer chromatography, we can see that paper sheets as such probably will be supplanted by thin-layer cellulose media. This means that the same immobile–mobile systems that are applied to paper can also be used for cellulose thin layer. The inorganic adsorbents do not appear to be very efficient for resolving highly polar compounds. When they are impregnated with immobile phases, ascension times are increased greatly because the capillaries are being

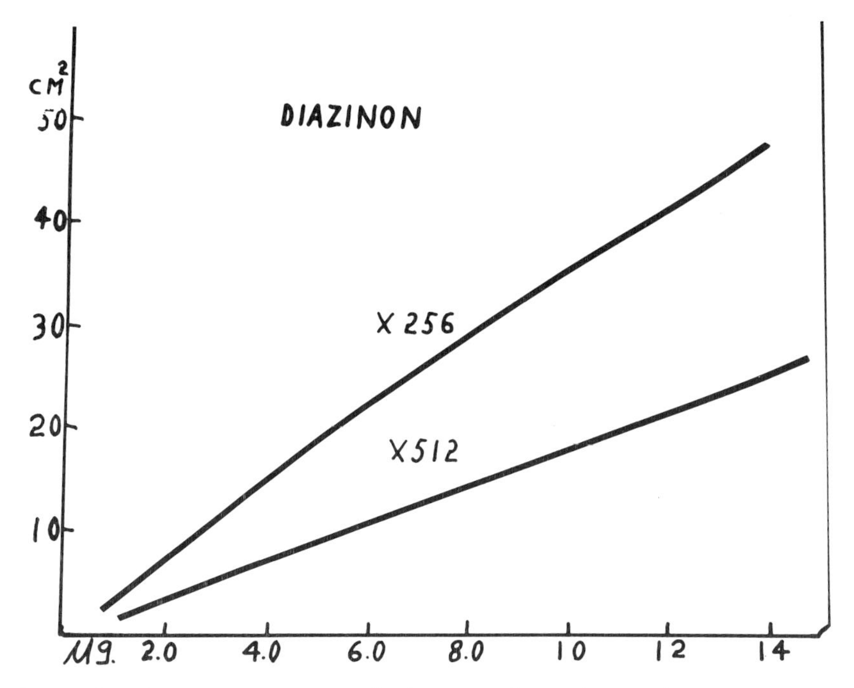

Figure 9. A comparison of standard curves obtained from diazinon at two different attenuation levels

filled with a liquid, and liquid–liquid partitioning is also taking place. In some instances, the ascension times are too long for practical purposes. With thin-layer cellulose, we can utilize the separations produced by immobile–mobile phases with better resolution because the spots produced are more compact than that obtainable from paper.

Any material with a capillary structure that can be made to adhere as a layer can be used for thin-layer chromatography. Quantitative determinations by this technique are bound to increase as more and more laboratories obtain scanning equipment. This will give impetus to the companies who make commercially-prepared plates to produce them with more care and uniformity.

Literature Cited

(1) Mrak, E. M., U. S. Department of Health, Education and Welfare, "Report on the Secretary's Commission on Pesticides and their Relationship to Environmental Health," Parts I and II, 1969.
(2) Morley, H. V., *Residue Rev.* (1966) **16,** 1.
(3) Samuel, B. L., Hodges, H. K., *Residue Rev.* (1967) **17,** 45.
(4) Thornburg, W. W., "Analytical Methods for Pesticides, Plant Growth Regulators and Food Additives," Vol. I, p. 87–108, Academic, New York, 1963.
(5) Biros, F., Burke, J. A., Gaul, J. A., Corneliussen, P. E., "Pesticide Analytical Manual," Vol. I, Ch. 4, 5, 6, U. S. Department of Health, Education and Welfare, Food and Drug Administration.
(6) Duggan, R. E., Barry, H. C., Johnson, L. Y., Williams, S., "Pesticide Analytical Manual, Vol. II, U. S. Department of Health, Education and Welfare, Food and Drug Administration.
(7) Consden, R., Gordon, A. H., Martin, J. P., *Biochem. J.* (1944) **38,** 224.
(8) Mitchell, L. C., *J. Assoc. Offic. Agr. Chemists* (1957) **40,** 999.
(9) Mitchell, L. C., *J. Assoc. Offic. Agr. Chemists* (1958) **41,** 781.
(10) Müller, R., Ernst, G., Schock, H., *Mitt. Gebiete Lebensm. Hyg.* (1957) **48,** 152.
(11) McKinley, W. P., Mahon, J. H., *J. Assoc. Offic. Agr. Chemists* (1959) **42,** 725.
(12) Mills, P., *J. Assoc. Offic. Agr. Chemists* (1959) **42,** 734.
(13) Bates, J. A., Jr., *Analyst* (1965) **90,** 453.
(14) Evans, W. H., *Analyst* (1962) **87,** 569.
(15) Getz, M. E., *Residue Rev.* (1963) **2,** 9.
(16) Major, A., Jr., Barry, H. C., *J. Assoc. Offic. Agr. Chemists* (1961) **44,** 202.
(17) Major, A., Jr., *J. Assoc. Offic. Agr. Chemists* (1962) **45,** 387.
(18) Rusiecki, W., Henneberg, M., Turkowska, T., *Biul. Inst. Ochrony Roslin* (1965) **32,** 23.
(19) Storherr, R. W., Tighe, J. F., Sykes, J. F., *J. Assoc. Offic. Agr. Chemists* (1960) **43,** 731.
(20) Storherr, R. W., Onley, J., *J. Assoc. Offic. Agr. Chemists* (1962) **45,** 382.
(21) Yip, G., *J. Assoc. Offic. Agr. Chemists* (1962) **45,** 367.
(22) Zweig, G., Archer, T. E., *J. Agr. Food Chem.* (1958) **6,** 910.
(23) Meinhard, J. E., Hall, N. J., *Anal. Chem.* (1949) **21,** 185.
(24) Kirchener, J. G., Heller, G. J., *Anal. Chem.* (1951) **23,** 420.
(25) Sola, T., Salimenen, K., Fiskari, K., *Z. Lebensm. Untersuch. Forsch.* (1962) **117,** 369.

(26) Walker, K. C., Beroza, M., *J. Assoc. Offic. Agr. Chemists* (1963) **46,** 250.
(27) Kovacs, M. J., Jr., *J. Assoc. Offic. Agr. Chemists* (1963) **46,** 884.
(28) Kovacs, M. J., Jr., *J. Assoc. Offic. Agr. Chemists* (1964) **47,** 1097.
(29) Conkin, R., *Residue Rev.* (1964) **6,** 136.
(30) Abbot, D. C., Thomson, J., *Residue Rev.* (1965) **11,** 1.
(31) Getz, M. E., *J. Assoc. Offic. Agr. Chemists* (1962) **45,** 393.
(32) Storherr, R. W., Getz, M. E., Watts, R. R., Friedman, S. J., Erwin, F., Giuffrida, L., Ives, F., *J. Assoc. Offic. Agr. Chemists* (1964) **47,** 1087.
(33) Beck, E. W., Johnson, J. C., Jr., Getz, M. E., Skinner, F. B., Dawsey, L. H., Woodham, D. W., Derbyshire, J. C., *J. Econ. Entomol.* (1968) **61,** 605.
(34) Beroza, M., Hill, K. R., Norris, K., *Anal. Chem.* (1968) **40,** 1608.
(35) Getz, M. E., "An Automatic Spotter for Quantitative TLC Analysis by Optical Scanning," Meeting of the Association of Official Analytical Chemists, Washington, D. C., October 1970.
(36) Getz, M. E., Friedman, S. J., *J. Assoc. Offic. Agr. Chemists* (1963) **46,** 707.
(37) Getz, M. E., Wheeler, H. G., *J. Assoc. Offic. Anal. Chemists* (1968) **51,** 1101.
(38) Beroza, M., Getz, M. E., Collier, C. W., *Bull. Environ. Contam. Toxicol.* (1968) **3,** 18.
(39) Adams, H. R., Schechter, M. S., Meeting of the Association of Official Agricultural Chemists, 77th, Washington, D. C., October 14–17, 1963.
(40) Watts, R. R., *J. Assoc. Offic. Agr. Chemists* (1963) **46,** 707.
(41) Shellard, E. J., Ed., "Quantitative Paper and Thin Layer Chromatography," p. 1–133, Academic, New York, 1968.
(42) Giddings, J. C., *J. Chromatog.* (1959) **2,** 48.

RECEIVED July 6, 1970. Mention of proprietary products is for identification only and does not imply endorsement of these products by the U. S. Department of Agriculture.

9

Applications of Combined Gas Chromatography–Mass Spectrometry to Pesticide Residue Identifications

FRANCIS J. BIROS

Perrine Primate Research Branch, Division of Pesticide Chemistry and Toxicology, Food and Drug Administration, U. S. Dept. of Health, Education, and Welfare, Perrine, Fla. 33157

Several individual and multiresidue analytical methods are available for gas chromatographic–mass spectrometric confirmation and identification of pesticide residues. Specific examples considered in this report include the analysis of intact and derivatized phenolic residues such as pentachlorophenol and 1-naphthyl chloroacetate, organophosphorus insecticide metabolic and hydrolytic products including O,O-*diethyl* O-*methyl phosphorothionate,* O,O-*diethyl* S-*methyl phosphorothiolate, and* O,O-*diethyl* O-*methyl phosphate, several organochlorine pesticides of the DDT and cyclodiene type, phenoxyalkanoic acid herbicide exposure and methodology studies involving 2,4-D and 2,4,5-T, and polychlorinated biphenyl residues. Conventional residue analytical methodology and gas chromatographic column technology are, in general, directly applicable to the analysis of human, animal, and environmental substrates by the combined technique. Evaluation of mass spectral fragmentation pathways provides definitive and conclusive confirmation of residue identity as well as characterization of residues and their metabolites of unknown structure.*

Unequivocal identification of pesticides, pesticide metabolites, and other chemical environmental pollutants in most cases requires more evidence than can be provided by a single chromatographic method. Several approaches may be utilized to secure firm proof of identity. For example, supportive chromatographic data such as relative retention values on two gas chromatographic columns of differing characteristics may be obtained. Of further value is information derived by means of thin-layer chromatography (R_f values), the solvent partitioning characteristics of the residue of interest (p-values), and the response behavior on several more or less specific gas chromatographic detectors. In addition, spectrometric techniques such as infrared and ultraviolet spectrophotometry may be applied to provide even less ambiguous proof of residue structural identity. The value of mass spectrometry as a tool for providing the structural identity of complex organic molecules has become well recognized in recent years. Mass spectra furnish information concerning the structural arrangement of atoms within a molecule on the basis of the mode of fragmentation of the compound as a radical ion, usually produced by electron bombardment. Fragmentation patterns resulting from bond fission and rearrangement of atoms are highly diagnostic and characteristic of the original molecular structure. Interpretations are usually made on the basis of empirical correlations, comparison with the mass spectra of standard materials, or detailed studies of ion decomposition mechanisms. The theories and principles governing the interpretation of mass spectra have been discussed in many reference texts and comprehensive reviews (*1*, *2*).

The potential utility of mass spectrometry in evaluating the metabolic pathways of pesticidal chemicals by providing the structural identity of metabolites was stated as early as 1962 in a review by Gunther (*3*). More recently, several reviews (*4*, *5*, *6*) have considered the role of mass spectrometry in chemical structure evaluations with special reference to pesticide residue analysis.

Correlations of mass spectral fragmentation pathways with the structure of pesticides are required to obtain background information useful in the interpretation of the mass spectra of unknown pesticide metabolites and other conversion products. Fortunately, the chemical modes of fragmentation of a large number of pesticidal compounds have been determined (*7*). This information is particularly useful in those instances when only microgram quantities of materials are available for characterization; mass spectrometry offers one of the best instrumental approaches in view of the low sensitivity of such complementary techniques as nuclear magnetic resonance spectrometry. The investigator may then rely on this single source of information for all the structural data. An additional advantage of the mass spectrometric approach is the availability of combined gas chromatograph–mass spectrometer instruments which permit high sensitivity analyses of multicomponent mixtures. Mass spectrometry, of course, has been widely employed as an individual technique for analysis of residues and metabolites isolated by conventional separation techniques such as thin-layer, liquid, paper, gas–liquid, and column chromatography. In these instances, specialized techniques are employed for the efficient trapping and collection of individual gas chromatographic peaks or thin-layer chromatographic spots for subsequent mass spectral analysis by standard sample vacuum introduction methods.

Recent reports have focused attention on the need for confirming pesticide residues, particularly those isolated from environmental samples (8, 9), and for chemical identity in ultramicroanalysis in general. Of the spectrometric techniques available for confirmation of residues, the combined gas chromatographic–mass spectrometric approach affords many advantages including relatively high sensitivity, elimination of the necessity for isolating minute quantities of pure samples with standard chromatographic methods, and a greater certainty in identification of an eluted component than that achieved by conventional detector response characteristics and gas chromatographic retention times alone. Because of the definitive structural information obtained by mass spectrometry, proper use of the technique would obviate the necessity for the application of two or more different carefully selected techniques for characterizing pesticide residues (8).

An additional obvious advantage would be the characterization and identification of frequently encountered, unknown components of extracts, which may represent pesticidal metabolites, photochemical and other "weathered" residues, nonpesticidal chemical pollutants, or co-extractive interferences whose identity may be required.

This report presents a discussion of recent applications of combined gas chromatography–mass spectrometry to analysis of pesticide residues isolated from human, animal, and environmental media. Emphasis will be placed on analytical techniques, related gas chromatographic column technology, and analysis of fragmentation pathways pertinent to the identification of pesticide residues.

Experimental

The mass spectrometric analyses initially reported in this communication were performed with a double focusing low resolution mass spectrometer; Model 270, Perkin-Elmer Corp., Norwalk, Conn., coupled through a Watson–Biemann type of molecular separator with a gas chromatographic system. (Commercial sources and trade names are provided for identification only. Their mention does not constitute endorsement by the Public Health Service or by the U. S. Dept. of Health, Education, and Welfare.)

Coiled glass gas chromatographic columns were employed except where otherwise indicated. Programmed temperature analyses were made with the initial and final oven conditions as well as the program rate given in the discussion. The molecular separator and gas inlet temperatures were maintained at 200° and 210°C, respectively. All mass spectra were recorded at 80 eV electron energy with 2400 V accelerating voltage; the filament emission current was 100 μa. Chromatograms were recorded from the total ion current monitor located between the electrostatic and magnetic analyzer sectors. Helium carrier gas was approximately 10 ml per min for the packed chromatographic columns and approximately 2 ml per min for the capillary columns. Injector temperature was maintained at 165°C. Mass spectra were scanned magnetically over the range of interest at a rate of either 3 or 10 sec per decade. The combined gas chromatograph–mass spectrometer system has been described in more detail elsewhere (*10*).

Phenolic Residues

Confirmation of the identity of pentachlorophenol (PCP) in samples of human blood, urine, tissue, clothing, and bedding materials has been reported (*11*), utilizing combined gas chromatographic–mass spectrometric analysis of hexane extracts of these substrates.

This report constituted a detailed account of the analytical aspects of an epidemic of infant deaths associated with the ingestion of this compound, which was the active ingredient of a mildew preventative, the suspected source of exposure. Analysis of the hexane extracts was performed on a 1.0 m × 2.5 mm coiled glass gas chromatographic column packed with 3% DEGS and 2% concentrated phosphoric acid. Under these conditions, analysis could be performed without prior derivatization of the PCP. All sample spectra gave four identical major ions when compared with PCP standard mass spectrum obtained at 20 eV. Thus, identity of the residues of PCP was confirmed on the basis of the observation of the molecular ion peak at m/e 264, a fragment ion of m/e 229, owing to elimination of a Cl atom, a major fragment resulting from the consecutive loss of HCl and CO at m/e 200, and finally a characteristic fragment of

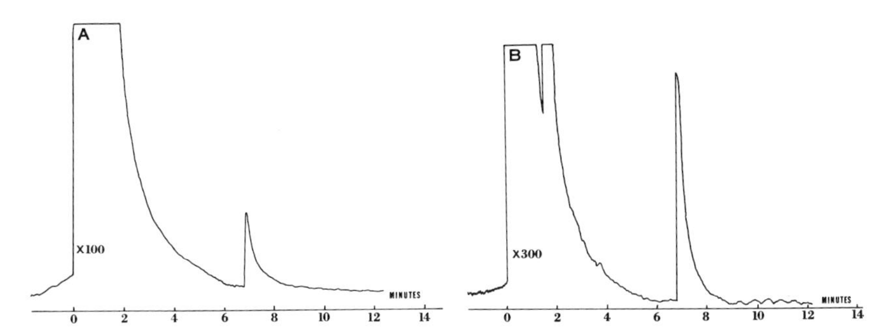

Figure 1. Total ion current chromatograms of (A) standard 1-naphthyl chloroacetate (1 μg) and (B) 1-naphthyl chloroacetate isolated from human urine
Programmed temperature conditions: two minutes at 165°C, to 185°C at 5°C/min, isothermal at 185°C

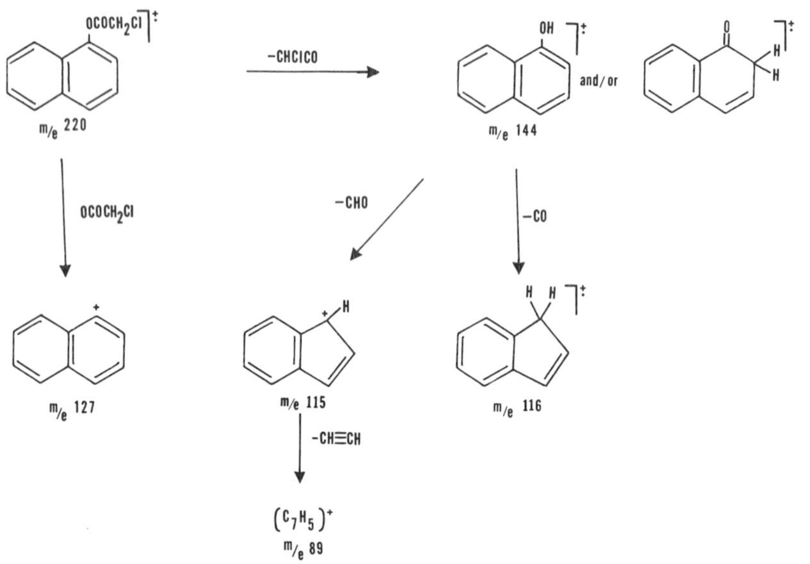

Figure 2. Electron impact-induced fragmentation scheme for 1-naphthyl chloroacetate

m/e 165 undoubtedly resulting from loss of a chlorine atom from the m/e 200 ion.

During the course of analytical method studies involving the development of procedures for the detection and quantitation of 1-naphthol, a major metabolite of carbaryl, in human and animal urine (*12*), the derivatized residue, 1-naphthyl chloroacetate, was confirmed by gas chromatography–mass spectrometry. The analytical procedure for the isolation of 1-naphthol residues involved acid hydrolysis of 1-naphthol conjugates in the urine, benzene extraction, derivatization with chloroacetic anhydride and pyridine, and removal of interfering material in urine by column chromatography using silica gel. 1-Naphthyl chloroacetate was eluted with 60% benzene–hexane. Programmed temperature gas chromatographic analysis with a coiled glass column, 4 ft $\times$ 1/8 inch o.d., packed with 2% SE-30 on 60/80 mesh Gas Chrom Q, was used to confirm residues of 1-naphthol as the chloroacetate ester in human urine of individuals occupationally exposed to carbaryl (Figure 1). Diagnostic mass spectral peaks observed for 1-naphthyl chloroacetate included the molecular ion of m/e 220, a base peak fragment which may be formulated as a 1-naphthol radical ion (or an isomer) found at m/e 144 arising from hydrogen rearrangement and loss of —CHClCO, a relatively weak peak at m/e 127 formed either by elimination of —$OCOCH_2Cl$ from the molecular ion and/or by loss of water from the m/e 145 fragment, and finally, other peaks characteristic of the fragmentation of 1-naphthol at m/e 115 and m/e 116 arising from expulsion of CHO and CO, respectively, and a peak at m/e 89 presumably resulting from a $C_7H_5^+$ ion formed by the loss of acetylene (C_2H_2) from the m/e 115 fragment (Figure 2).

Organophosphorus Pesticides

In connection with the development of an analytical method (*13*) for the determination of organophosphorus pesticides in human blood and urine, mass spectral confirmation of a series of methylated and ethylated derivatives of the hydrolytic and metabolic products of these insecticides was required. The urine of an individual occupationally exposed to parathion was extracted with a 1:1 (v/v) solvent mixture of acetonitrile and diethyl ether. Simultaneously, the intact organophosphorus insecticides were hydrolyzed by adding a portion of 5*N* hydrochloric acid to

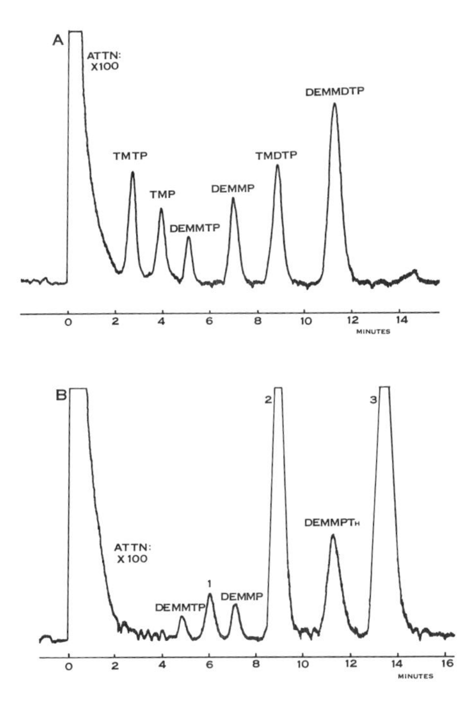

Figure 3. Total ion current chromatograms of (A) standard mixture of methylated dialkylphosphates, phosphorothioates, and phosphorodithioates (2 μg) and (B) human urine extract containing methylated hydrolytic and metabolic products of parathion

Programmed temperature conditions: five minutes at 75°C, to 120°C at 5°C/min, isothermal at 120°C

the urine. Following extraction, the organic layer was concentrated and the residues in solution derivatized by means of diazomethane. The urine extract was then subjected to a clean-up procedure employing silica gel column chromatography (*14*), and benzene (I), 40% ethyl acetate in benzene (II), and 80% ethyl acetate in benzene (III) as eluting solvents. Concentration of the chromatographic column eluents provided the urine extract which was subjected to programmed temperature analyses on a coiled glass capillary column, 110 ft $\times$ 0.025 inch i.d., coated with Versamid 900 containing 5% Igepal CO-880. The resulting total ion current chromatogram is shown in Figure 3. Also illustrated is the total ion current trace obtained by analysis of a series of methylated dialkyl phosphate standard materials.

The metabolic and/or hydrolytic products of parathion encountered as residues in the urine include both diethyl phosphoric acid and diethyl phosphorothioic acid, most probably as their salts (potassium or sodium). Derivatization of these residues with diazomethane would result in the formation of three trialkyl phosphate compounds, namely, *O,O*-diethyl *O*-methyl phosphate (DEMMP), *O,O*-diethyl *O*-methyl phosphorothionate (DEMMTP), and *O,O*-diethyl *S*-methyl phosphorothiolate (DEMMPTh). Earlier (*15*), it had been shown by combined gas chromatography–mass spectrometry and other analytical data that a later-eluting major product (*ca.* 85%) of the methylation of diethyl phosphorothioic acid formed under the conditions of the analytical method was DEMMPTh, and the minor product formed (*ca.* 15%) was DEMMTP. Accordingly, all three trialkyl phosphates were observed and confirmed by mass spectrometry in the analysis of the human urine extract. Sufficient internal bond energy differences are associated with the isomeric structures DEMMPTh and DEMMTP that qualitatively and quantitatively dissimilar fragmentation patterns are observed for both isomers as can be seen from the mass spectra of these compounds shown in Figure 4.

The mass spectrum of DEMMTP is characterized by both phosphorus–oxygen and carbon–oxygen bond fission as evidenced by the formation of fragment ions at m/e 140 ($M-C_2H_4O$) and m/e 156 ($M-C_2H_4$). Single and double hydrogen rearrangements accompany the loss of the ethyl substituent. Subsequent fragmentation reactions noted were similar bond fission involving elimination of the remaining alkyl substituents as an ethylene molecule or an ethoxyl radical. A relatively intense peak, for example, of m/e 111 [$CH_3OP(:S)-OH^+$], appears to be formed by consecutive loss of ethylene (to form the m/e 156 ion) and an ethoxyl substituent from the molecular ion as evidenced by metastable peaks at m/e 132.26 and m/e 78.98 observed for these two processes. Significant fragment ions were also formed through loss of both sulfur atoms or sulfhydryl

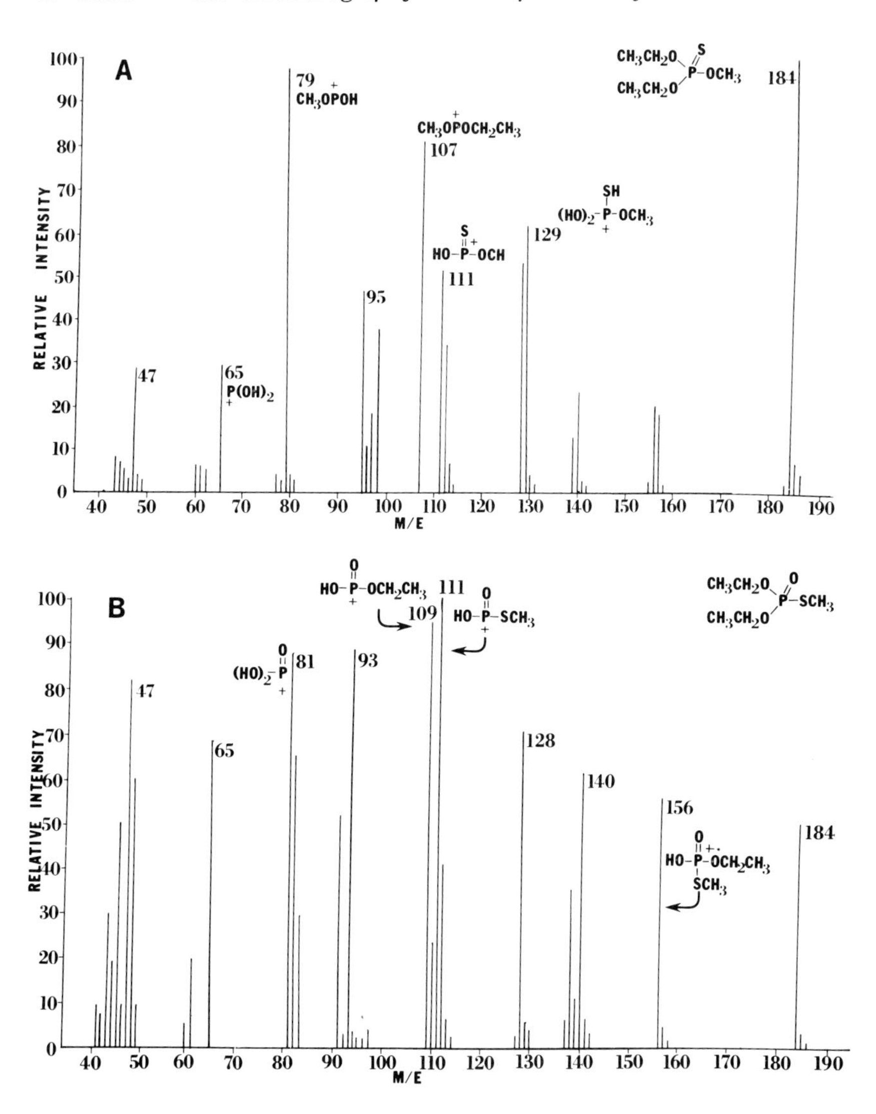

Figure 4. Mass spectra of (A) O,O-*diethyl* O-*methyl phosphorothionate (DEMMTP), and (B)* O,O-*diethyl* S-*methyl phosphorothiolate (DEMMPTh)*

radicals from intermediate daughter ions. Figure 5 illustrates the major dissociative processes observed in the mass spectrum of DEMMTP. The P(:S)–OR to P(:O)–SR isomerization induced under electron impact conditions for compounds of this structure could not be confirmed for DEMMTP although other investigators have observed this rearrangement for organic phosphorothionates (*16, 17*). Other trialkyl phosphorothionate

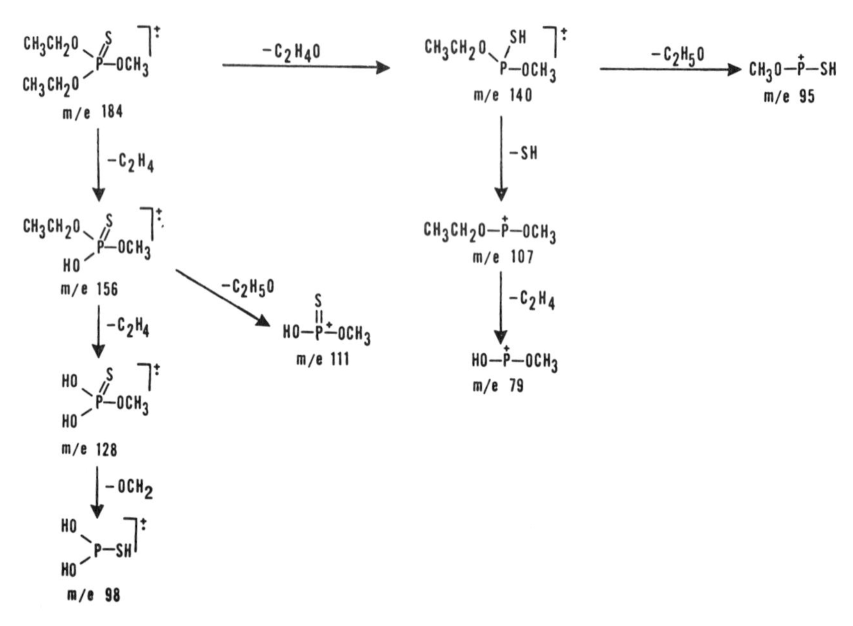

Figure 5. Electron impact-induced fragmentation scheme for O,O*-diethyl* O*-methyl phosphorothionate (DEMMTP)*

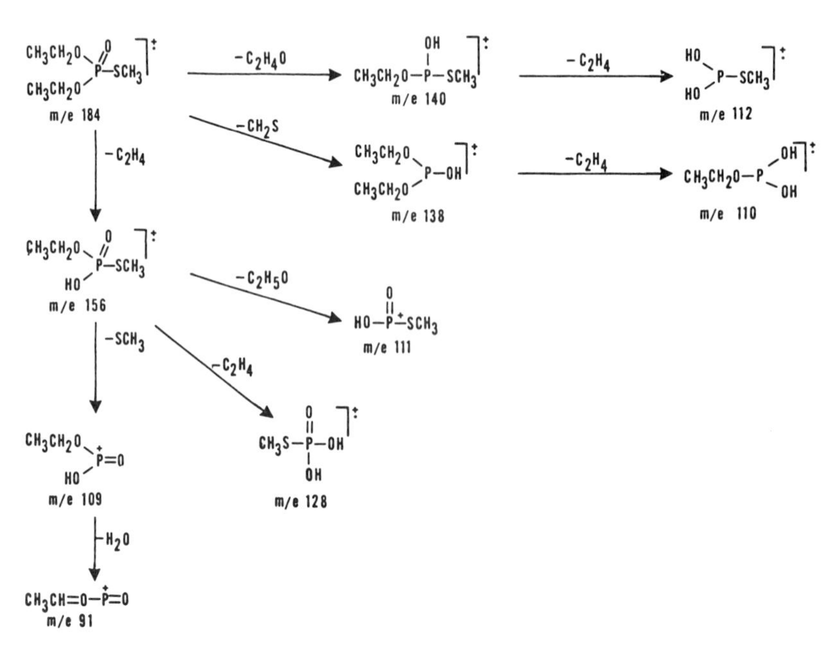

Figure 6. Electron impact-induced fragmentation scheme for O,O*-diethyl* S*-methyl phosphorothiolate (DEMMPTh)*

compounds have also been observed to undergo this rearrangement process (*18*).

A somewhat more complex dissociative scheme was observed for DEMMPTh (Figure 6). The molecular ion of this compound decomposes by fission of a carbon–oxygen bond (m/e 156, M–C_2H_4), phosphorus–oxygen bond (m/e 140, M–C_2H_4O), and the phosphorus–sulfur bond (m/e 138, M–CH_2S). The structure of the m/e 156 fragment (Figure 6) permits a subsequent identical triple decomposition scheme to yield fragments of m/e 128 (156–C_2H_4), m/e 109 (156–CH_3S), and m/e 111 (156–C_2H_5O). Another significant reaction observed was the elimination of water which, for example, led to the formation of an intense ion fragment at m/e 91 (109–H_2O), characteristic of the electron impact fragmentation of esters.

The decomposition of DEMMP under electron impact results in the formation of significant fragment ions of m/e 141, m/e 113, and m/e 95, according to the dissociative scheme illustrated in Figure 7.

Organochlorine Pesticides

Chlorinated hydrocarbon pesticide residues in human adipose tissue and liver tissue samples have been identified by mass spectrometry coupled with gas chromatography. A general, extensive extraction and cleanup procedure adapted from existing methods was used to isolate

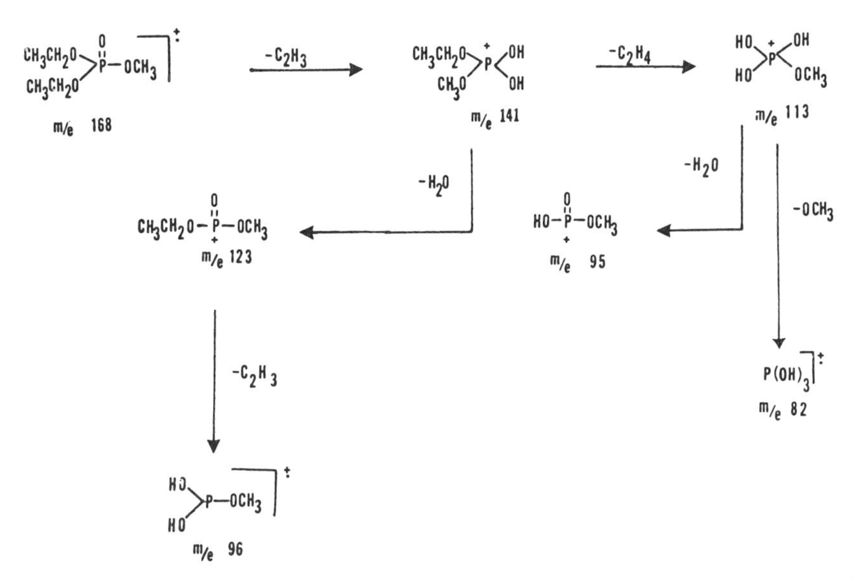

Figure 7. Electron impact-induced fragmentation scheme for O,O-diethyl O-methyl phosphate (DEMMP)

and purify the organochlorine residues. The tissue samples were extracted with acetonitrile or petroleum ether, followed by liquid–liquid partitioning of the extract between hexane or petroleum ether and acetonitrile. Column chromatography and cleanup on Florisil provided two individual fractions (*19*), which were then concentrated and subjected to further cleanup and pesticide residue fractionation by means of silica gel column chromatography employing a published procedure (*20*). The pesticide residues were separated by the silica gel chromatographic step into two fractions with sufficient resolution on the gas chromatographic column for the identification of seven pesticide residues from the adipose tissue sample and six pesticide residues from the liver tissue sample. These included: β- and γ-hexachlorocyclohexane (HCH), 1,4,5,6,7,8,8-heptachloro-2,3-epoxy-2,3,3a,4,7,7a-hexahydro-4,7-methanoindene (heptachlor epoxide), 1,2,3,4,10,10-hexachloro-6,7-epoxy-1,4,4a,5,6,7,8,8a-octahydro-1,4-*endo,exo*-5,7-dimethanonaphthalene (dieldrin), 2,2-bis(*p*-chlorophenyl)-1,1-dichloroethylene (*p,p'*-DDE), 2,2-bis(*p*-chlorophenyl)-1,1-dichloroethane (*p,p'*-DDD), and 1,1-bis(*p*-chlorophenyl)-2,2,2-trichloroethane and 1-(*o*-chlorophenyl)-1-(*p*-chlorophenyl)-2,2,2-trichloroethane (*p,p'*-DDT and *o,p'*-DDT, respectively). The concentration of the pesticides in tissue ranged from 0.073 to 28.7 ppm. Gas chromatographic separations were performed on a 4 ft × 1/8 inch o.d. coiled glass column packed with 3% OV-210 on 80/100 mesh Chromosorb W (HP). Figure 8 illustrates the total ion current chromatogram obtained by analysis of the adipose tissue extract representing 5 grams of tissue.

Instrumental sensitivity was sufficient to identify residues at concentrations of 0.30 ppm. Computer enhancement techniques (*21*) permitted identification of heptachlor epoxide residues at a level of 0.073 ppm.

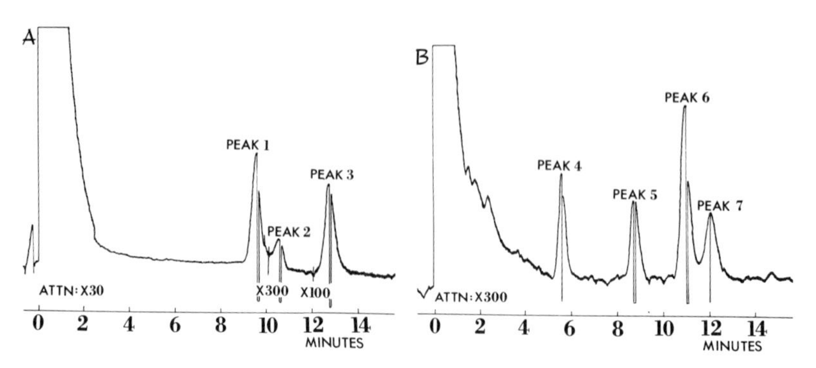

Figure 8. Total ion current chromatograms of extracts containing organochlorine pesticide residues isolated from adipose tissue (A) fraction 1 and (B) fraction 2

Programmed temperature conditions: one minute at 150°C, to 200°C at 5°C/min, isothermal at 200°C

Definitive confirmation of pesticide residues was obtained by comparison of parent and fragment ion intensities and mass numbers of eluted pesticides and reference pesticides. Table I lists the residues encountered and the mass numbers and intensities of the characteristic fragments employed for identification in the adipose tissue sample. The mass spectral fragmentation patterns for all the compounds included in Table I with the exception of β-HCH have been adequately discussed by other investigators (7).

Table I. Gas Chromatographic Peak Identities and Characteristic Mass Spectral Fragments and Intensities of Some Organochlorine Pesticide Residues Isolated From Human Adipose Tissue

Peak No.	*Pesticide*	*Characteristic Mass Spectral Peaks and Intensities*
1	*p,p'*-DDE	m/e 246 (100%), 318 (83%), 316 (Molecular ion, 66%), 248 (58%), 320 (41%), 176 (41%), 210 (16%).
2	*o,p'*-DDT	m/e 235 (100%), 237 (70%), 165 (40%), 75 (20%), 199 (18%), 246 (18%), 352 (Molecular ion, 3%).
3	*p,p'*-DDT	m/e 235 (100%), 237 (72%), 165 (48%), 75 (22%), 50 (18%), 51 (9%), 352 (Molecular ion, 2%).
4	β-HCH	m/e 109 (100%), 219 (85%), 181 (78%), 183 (80%), 111 (62%), 193 (62%), 288 (Molecular ion, 4%).
5	Heptachlor epoxide	m/e 81 (100%), 353 (84%), 355 (76%), 351 (48%), 357 (35%), 237 (33%), 386 (Molecular ion, 8%).
6	Dieldrin	m/e 79 (100%), 108 (19%), 263 (18%), 277 (19%), 279 (16%), 345 (7%), 378 (Molecular ion, 3%).
7	*p,p'*-DDD	m/e 235 (100%), 237 (66%), 165 (58%), 75 (21%), 82 (16%), 88 (16%), 318 (Molecular ion, 5%).

Some difficulties were encountered in the absolute confirmation of configurational and structural isomers where mass spectral peak intensities and mass values of characteristic fragment ions were too similar under the conditions of the analysis. In those instances, however, identification was readily made on the basis of gas chromatographic retention time data. The estimated lower limit of detectability of organochlorine pesticide residues by this analytical scheme and instrument configuration was determined to be approximately 0.05 to 0.1 ppm.

Herbicide Residues

Combined methodology and metabolism studies involving the herbicides 2,4-dichlorophenoxyacetic acid (2,4-D) and 2,4,5-trichlorophenoxyacetic acid (2,4,5-T) furnished an opportunity for the application of the gas chromatographic–mass spectrometric technique to the confirmation of derivatized intact residues and metabolites and the structural characterization of unknown metabolites isolated from rat urine. The analytical technique, briefly summarized, involves acid hydrolysis of phenolic conjugates, extraction of free phenols and acids with benzene, extract concentration and derivatization with diazoethane, and finally, column chromatographic cleanup and separation by silica gel chromatography employing benzene–hexane and benzene–ethyl acetate solvent systems as eluents (22). Programmed temperature gas chromatographic analyses of the urine extracts with a 7 ft × 1/8 inch o.d. aluminum column packed with 20% OV-101 on 60/80 mesh Gas Chrom Q provided the total ion current chromatograms shown in Figure 9. The mass spectral fragmentation data were used to confirm the presence of 2,4-D and 2,4,5-T as the ethyl esters and trichlorophenol (TCP) as the ethyl ether in the several

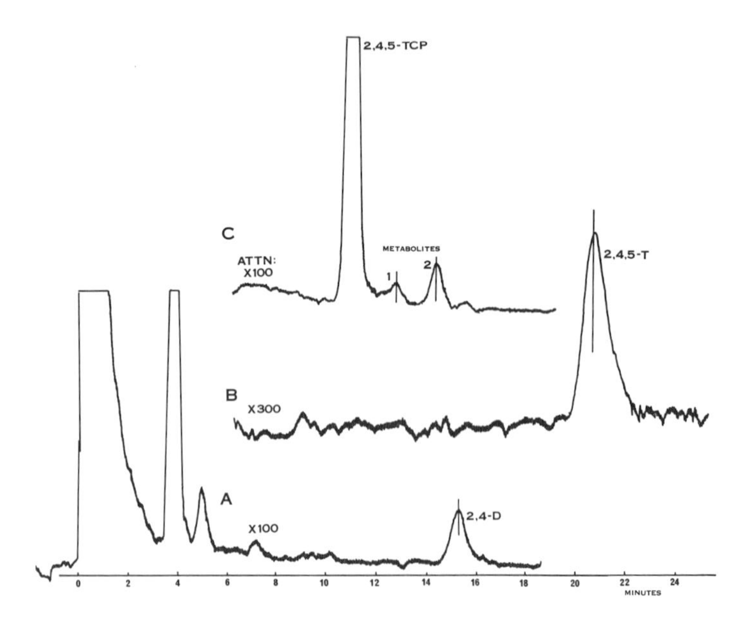

Figure 9. Total ion current chromatograms of animal urine extracts containing (A) 2,4-D, (B) 2,4,5-T, and (C) 2,4,5-trichlorophenol and an isomer of trichlorodihydroxybenzene, peak 2 (as ethyl esters and mono- and diethyl ethers, respectively)
Programmed temperature conditions: five minutes at 180°C, to 220°C at 5°C/min, isothermal at 220°C

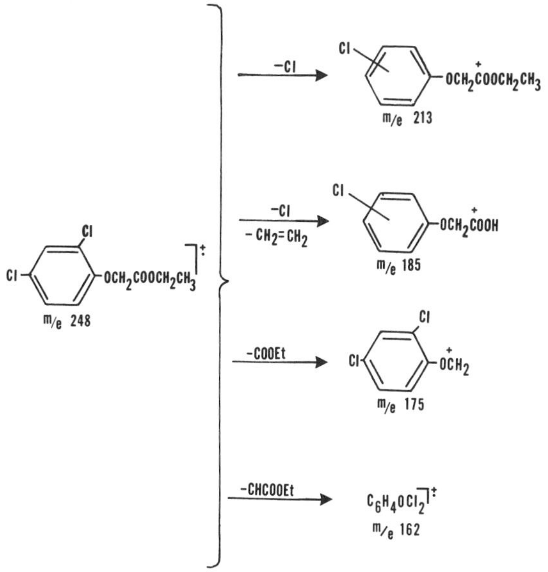

Figure 10. Electron impact-induced fragmentation scheme for 2,4-dichlorophenoxyacetic acid, ethyl ester

analyses performed. The diagnostic fragmentation pattern for 2,4-D ethyl ester included peaks at m/e 248, molecular ion; m/e 213, elimination of a Cl atom; m/e 185, consecutive elimination of a Cl atom and ethylene (C_2H_4) from the parent ion species; m/e 175, loss of the carboethoxyl function from the molecular ion; and finally, a peak at m/e 162 presumably arising from a dichlorophenol or dichlorocyclohexadienone-type radical ion (Figure 10). An analogous fragmentation scheme was observed for the ethyl ester of 2,4,5-T: m/e 282, molecular ion; m/e 247, loss of chlorine atom; m/e 219, consecutive elimination of a chlorine atom and a molecule of ethylene from the parent ion; m/e 209, decomposition of the parent ion with elimination of the carboethoxyl group; and finally, m/e 196, corresponding to a trichlorophenol or trichlorocyclohexadienone-type radical ion species.

In addition to the intact residues of 2,4,5-T excreted in the rat urine, mass spectral evidence was obtained for the presence of the metabolite TCP and a trichlorodihydroxybenzene isomer, which were observed as their mono- and diethyl ether derivatives, respectively. Thus, TCP ethyl ether displayed a molecular ion at m/e 224 and characteristic fragments at m/e 196 and m/e 160 owing to consecutive elimination of ethylene and HCl from the parent ion. The fragmentation pattern of 2,4,5-trichloropenol was observed to exhibit similar behavior below m/e 196.

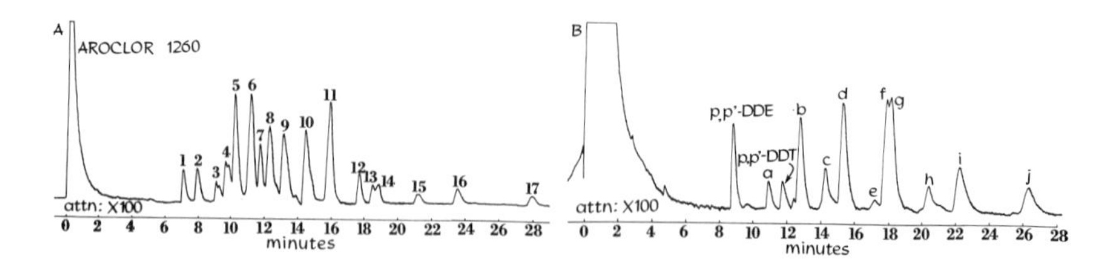

Figure 11. Total ion current chromatograms of (A) standard Aroclor 1260 mixture of polychlorinated biphenyls. Programmed temperature analysis: two minutes at 200°C, to 230°C/min, isothermal at 230°C. (B) Human adipose tissue extract. Programmed temperature analysis: two minutes at 190°C, to 230°C at 5°C/min, isothermal at 230°C.

The mass spectral evidence for the trichlorodihydroxybenzene diethyl ether included a molecular ion of m/e 286 and significant fragment ions at m/e 240, m/e 212, and m/e 176 arising from the consecutive elimination of two molecules of ethylene and one molecule of HCl.

Polychlorinated Biphenyls

Confirmatory evidence for the presence of polychlorinated biphenyls (PCB) in fish, seabirds, conifer needles, and human depot fat has been obtained by gas chromatography–mass spectrometry (*23, 24*). The mass spectral behavior of standards of PCB materials (Aroclor 1254 and 1260) and PCB compounds isolated from human adipose tissue utilizing standard analytical methodology (*19*) has been reported recently (*25*). An illustration of the total ion current chromatogram obtained by programmed temperature analysis of a mixture of standard PCB compounds (Aroclor 1260) on a 100 ft × 0.02 inch i.d. stainless steel capillary column coated with OV-1 silicone oil is depicted in Figure 11. Most components of the Aroclor standard produced mass spectra with molecular ion groups of high intensity. In addition, the characteristic isotopic distribution patterns corresponding to the number of chlorine atoms in the parent ion and chlorine-containing fragment ions were observed. Other noteworthy features of the mass spectra were the relatively intense ion fragments resulting from consecutive loss of chlorine atoms from the parent ion and the presence of intense doubly charged fragments within the mass spectra of most of the PCB compounds. Thus, it was possible to establish the molecular weight and number of chlorine atoms associated with each major numbered component in the chromatogram of Figure 11. The standard Aroclor 1260, on this basis, was shown to consist of at least two pentachlorobiphenyls, four hexachlorobiphenyls, six heptachlorobiphenyls, two octachlorobiphenyls, two nonachlorobiphenyls, and decachlorobiphenyl.

Although it was apparent that several of the major peaks in the chromatogram of the standard PCB material were in fact composed of more than a single component, no further attempt was made to elaborate on the identity or nonidentity of the individual components of each of these peaks.

Structural studies of the PCB components must necessarily involve the application of complex separation procedures coupled with detailed spectrometric studies of each isolated component. Figure 11 also includes the total ion current chromatogram of a human adipose tissue extract which was shown to contain traces of PCB compounds (peaks *a* through *j*) whose mass spectra were consistent with those found for Aroclor 1260, namely, peaks 6 and 9 through 17.

PCB materials in field-collected bald eagle specimens have been identified utilizing combined gas chromatography–mass spectrometry techniques in conjunction with thin-layer chromatographic separations of standard materials as well as tissue extracts (*26*). A total of 19 PCB compounds were detected in the field samples. Analyses were performed employing a spiral glass gas chromatographic column, 9 ft $\times$ 0.25 inch o.d., packed with 1% SE-30 on 100/120 mesh Gas Chrom Q. Mass spectra of the major components of Aroclor 1254 PCB standard were presented in this report as well as relative retention data for the individual components.

Discussion

Of all the systems which have been utilized for the analysis of pesticide residues, combined gas chromatography–mass spectrometry affords a particularly useful approach because positive identification of the components of a mixture can be made without prior separation at sensitivities compatible with the limited quantities of residues generally available. From the results of recent studies involving the application of this technique, it has been demonstrated that available residue analytical methods provide efficient isolation and adequate cleanup of extracts of human, animal, and environmental media in most cases to permit gas chromatographic–mass spectrometric analyses with maximum confidence. Additionally, it has been shown that this combined technique will conveniently provide definitive and conclusive confirmation of residue identity as well as characterization of residues and their metabolites of unknown structure.

In addition to the applications reported here, gas chromatography–mass spectrometry has been employed in the characterization of photochemical degradation products of *p,p′*-DDT and *p,p′*-DDE (*27*), synthetic intermediates in *p,p′*-DDT metabolism studies (*28*), and transformation products of herbicidal chloroanilines in soil (*29*).

From a technique standpoint, additional studies should be made on improvements of instrumental sensitivity to allow routine confirmation of pesticide residues at the nanogram and possibly the picogram level. Some gains have already been reported in this area, utilizing computer enhancement techniques (*21*), pulse height analysis (*30, 31*), and other methods (*4*). Studies documenting compound yield and enrichment factors for the several classes of organic pesticides and various gas chromatographic–mass spectrometric interfacial systems under a variety of operating conditions are required. Technological studies systematically designed to exploit the advantages of capillary and other gas chromatographic columns for pesticide residue analysis are also needed. Finally, development of a complete residue analysis system employing either an electron capture or flame ionization detector and splitter arrangement in tandem with the molecular separator permitting simultaneous mass spectrometric identification and detection and quantitation of pesticide residues from a single gas chromatographic injection would offer obvious advantages.

The reported applications of combined gas chromatography–mass spectrometry to the analysis of pesticide residues have been increasing in frequency in recent years. With developments in instrument technology and availability of instrumentation, it can be expected that the technique will be relied upon to an even greater extent in the future.

Literature Cited

(1) Biemann, K., "Mass Spectrometry: Organic Chemical Applications," 1st ed., p. 46 ff., McGraw-Hill, New York, 1962.
(2) Budzikiewicz, H., Djerassi, C., Williams, D. H., "Mass Spectrometry of Organic Compounds," p. 6 ff., Holden-Day, San Francisco, 1967.
(3) Gunther, F. A., "Instrumentation in Pesticide Residue Determinations," *Advan. Pest Control Res.* (1962) **5,** 191–319.
(4) Biros, F. J., "Recent Applications of Mass Spectrometry and Combined Gas Chromatography–Mass Spectrometry to Pesticide Residue Analysis," *Residue Rev.* (1971) in press.
(5) Kantner, T. R., Mumma, R. O., "Application of Mass Spectroscopy to Pesticide Residue Analysis," *Residue Rev.* (1966) **16,** 138–51.
(6) Westlake, W. E., Gunther, F. A., "Advances in Gas Chromatographic Detectors Illustrated from Applications to Pesticide Residue Evaluations," *Residue Rev.* (1967) **18,** 175–217.
(7) Sphon, J. A., Damico, J. N., "The Mass Spectra of Some Chlorinated Aromatic Pesticidal Compounds," *Org. Mass Spectr.* (1970) **3,** 51–62 and references cited therein.
(8) Robinson, J., Richardson, A., Elgar, K. E., "Chemical Identity in Ultramicroanalysis," *152nd National Meeting, ACS, New York, September 1966,* paper #A-075.
(9) Schechter, M. S., "The Need for Confirmation," *Pestic. Monit. J.* (1968) **2,** 1.

(10) Biros, F. J., Walker, A. C., "Pesticide Residue Analysis in Human Tissue by Combined Gas Chromatography–Mass Spectrometry," *J. Agr. Food Chem.* (1970) **18,** 425–9.

(11) Barthel, W. F., Curley, A., Thrasher, C. L., Sedlak, V. A., "Determination of Pentachlorophenol in Blood, Urine, Tissue, and Clothing," *J. Assoc. Offic. Anal. Chemists* (1969) **52,** 294–8.

(12) Shafik, M. T., Sullivan, H., Enos, H. F., "A Method for the Determination of 1-Naphthol in Urine," *Bull. Environ. Contam. Toxicol.* (1971) **6,** 34–9.

(13) Shafik, M. T., Enos, H. F., "Determination of Metabolic and Hydrolytic Products of Organophosphorus Pesticide Chemicals in Human Blood and Urine," *J. Agr. Food Chem.* (1969) **17,** 1186–9.

(14) Shafik, M. T., Bradway, D., "A Cleanup Procedure for the Determination of Low Levels of Alkyl Phosphates, Thiophosphates, and Dithiophosphates in Rat Urine," *Joint ACS–Chem. Inst. Canada Intern. Meeting, Toronto, May 1970,* paper #PEST 054.

(15) Shafik, M. T., Bradway, D., Biros, F. J., Enos, H. F., "Characterization of Alkylation Products of Diethyl Phosphorothioate," *J. Agr. Food Chem.* (1970) **18,** 1174–5.

(16) Cooks, R. G., Gerrard, A. F., "Electron Impact Induced Rearrangements in Compounds Having the P=S Bond," *J. Chem. Soc.* B (1968) 1327–33.

(17) Jorg, J. R., Houriet, R., Spiteller, G., "Massenspektren von Pflanzenschutzmitteln," *Monatsh. Chem.* (1966) **97,** 1064–87.

(18) Biros, F. J., Ross, R. T., "Fragmentation Processes in the Mass Spectra of Trialkylphosphates, Phosphorothionates, Phosphorothiolates, and Phosphorodithioates," *18th Conf. Mass Spectr. Allied Topics, San Francisco, June 1970,* paper #G3.

(19) Food and Drug Administration, "Pesticide Analytical Manual," Vol. I, Sec. 211, *General Methods for Fatty Foods,* U.S. Dept. of Health, Education, and Welfare, Washington, D. C., revised July 1969.

(20) Kadoum, A. M., "Application of the Rapid Micromethod of Sample Cleanup for Gas Chromatographic Analysis of Common Organic Pesticides in Ground Water, Soil, Plant, and Animal Extracts," *Bull. Environ. Contam. Toxicol.* (1968) **3,** 65–70.

(21) Biros, F. J., "Enhancement of Mass Spectral Data by Means of a Time Averaging Computer," *Anal. Chem.* (1970) **42,** 537–40.

(22) Shafik, M. T., Sullivan, H., "A Method for the Determination of Low Level Exposure to 2,4-D and 2,4,5-T," *7th Inter-Am. Conf. Toxicol. Occupational Med., Miami, August 1970; J. Environ. Anal. Chem.* (1971) in press.

(23) Koeman, J. H., ten Noever de Brauw, M. C., de Vos, R. H., "Chlorinated Biphenyls in Fish, Mussels, and Birds from the River Rhine and the Netherlands Coastal Area," *Nature* (1969) **221,** 1126–8.

(24) Widmark, G., "Possible Interference by Chlorinated Biphenyls," *J. Assoc. Offic. Anal. Chemists* (1967) **50,** 1069.

(25) Biros, F. J., Walker, A. C., Medbery, A., "Polychlorinated Biphenyls in Human Adipose Tissue," *Bull. Environ. Contam. Toxicol.* (1970) **5,** 317–23.

(26) Bagley, G. E., Reichel, W. L., Cromartie, E., "Identification of Polychlorinated Biphenyls in Two Bald Eagles by Combined Gas Liquid Chromatography–Mass Spectrometry," *J. Assoc. Offic. Anal. Chemists* (1970) **53,** 251–61.

(27) Plimmer, J. R., Klingebiel, U. I., Hummer, B. E., "Photo-oxidation of DDT and DDE," *Science* (1970) **167,** 67–9.

(28) McKinney, J. D., Boozer, E. L., Hopkins, H. P., Suggs, J. E., "Synthesis and Reactions of a Proposed DDT Metabolite, 2,2-bis(*p*-Chlorophenyl) Acetaldehyde," *Experientia* (1969) **25,** 897–8.
(29) Kearney, P. C., Plimmer, J. R., Guardia, F. B., "Mixed Chloroazobenzene Formation in Soil," *J. Agr. Food Chem.* (1969) **17,** 1418–9.
(30) Bergstedt, L., Widmark, G., "Repetitive Scanning in Gas Chromatography–Mass Spectrometry," *Chromatographia* (1970) **3,** 59–63.
(31) Widmark, G., Bergstedt, L., Laveskog, A., "Mass Spectrometry of Organochlorine Compounds and Attempts to Reach Positive Identification," *I.U.P.A.C. Conference, Appl. Chem. Div., Sittingbourne, Kent, U.K., October 1968.*

RECEIVED September 3, 1970.

10

The Identification of Pesticides at Residue Concentrations

K. E. ELGAR

Shell Research Ltd., Woodstock Agricultural Research Centre,
Sittingbourne, Kent, England

It is important to confirm the identity of pesticide residues convincingly. Some methods, such as TLC, paper chromatography, or p-*values share the same physical property of partition in achieving separations of mixtures. They do not give independent evidence for the identity of a compound. Similarly, GLC retention times for a compound on different stationary phases are often highly correlated. Thus, the choice of confirmatory techniques should be carefully made. Although powerful methods such as GC/MS are being studied, there is a need for simpler operations—for instance, the formation of chemical derivatives. Experiments with aldrin and dieldrin have revealed a number of reactions which are convenient for the confirmation of residues of these compounds.*

The implications of finding a pesticide residue in a sample can often be far-reaching, and it is important that the identification of the residue is convincing (*1, 2, 3*). The authenticity of a trace constituent can be established by many different methods, some of these contributing much more evidence than others. If there is sufficient residue and it can be isolated in a reasonably pure state, the methods of infrared spectroscopy (*4*) and mass spectroscopy (*5, 6*) are very powerful tools and often give results which the analyst feels settle the question of identity. It is rare to be able to settle the question so easily. More often, the analyst is faced with a possible residue present at a very low concentration indeed, and he has to resort to other methods, the results of which are more ambiguous. It is important to choose analytical procedures which give independent evidence for the identity of the residue.

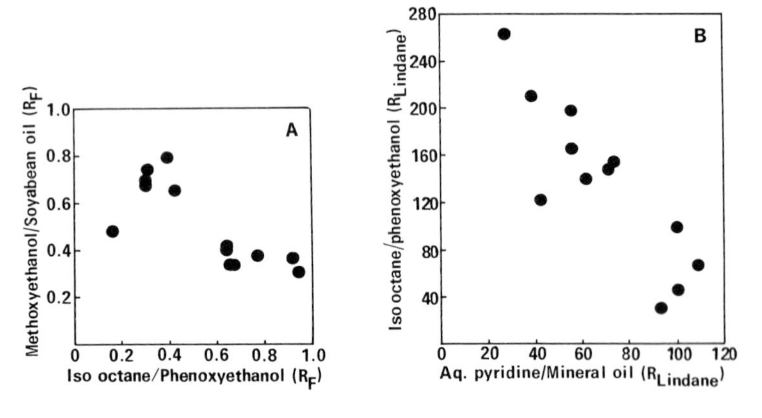

Journal of the Association of Official Agricultural Chemists

Figure 1. Paper chromatography; A) Ref. 7; B) Ref. 8

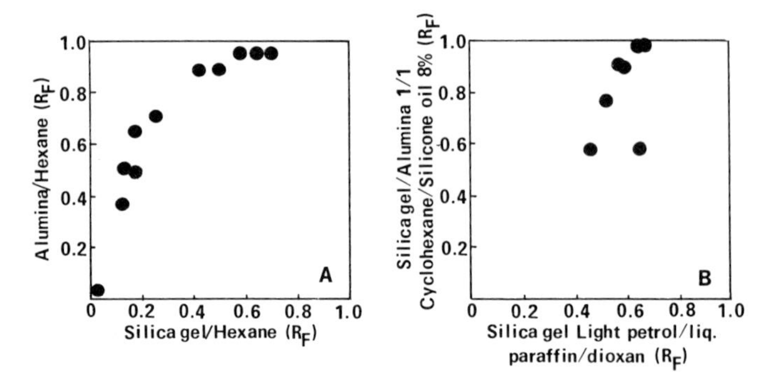

Figure 2. Thin-layer chromatography; Ref. 9

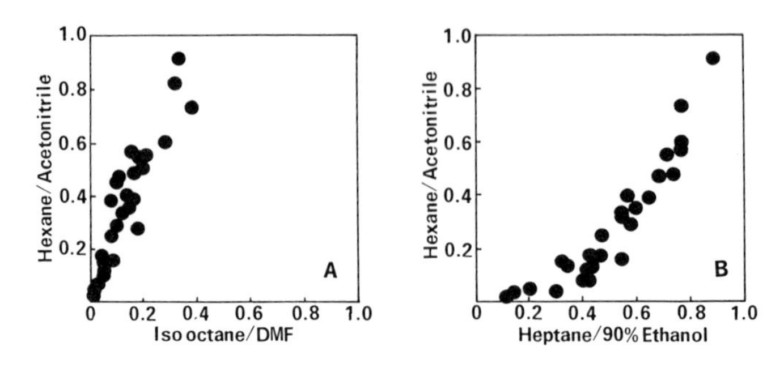

Journal of the Association of Official Agricultural Chemists

Figure 3. Extraction p-*values; Ref. 10*

In Figure 1, R_F values for organochlorine insecticides in different systems have been plotted (*7, 8*). It is clear that in general the points lie close to a line, and therefore there is a high degree of correspondence between the R_F of a compound in one system and the R_F in a completely different system. This shows that in order to confirm the identity of a residue it is not enough to run the extract in two or even several paper chromatography systems, because they all rely on one parameter—the partition coefficient of the compound between immiscible liquids.

In Figure 2, the R_F values of organochlorine insecticides in different thin-layer chromatography systems are shown (*9*). Again, it is evident that running portions of extract in different systems tells the analyst little more than running it in one system only.

In Figure 3, Bowman and Beroza's extraction *p*-values obtained from different pairs of immiscible solvents are plotted (*10*). The degree of correspondence for the *p*-values of these compounds—organochlorine and organophosphorus insecticides and other pesticides—from one pair of solvents to another is clearly featured. The additional independent evidence for identity gained by obtaining a second *p*-value is very small.

Figure 4 shows the recent data given by Thompson *et al.* (*11*). GLC retention times with the nonpolar dimethyl siloxane polymer DC 200 as the stationary phase are plotted against retention times on other phases which gave good resolution of individual pesticides. The correspondence between retention times on different phases is highly significant. With the stationary phases plotted in Figure 5, the degree of concordance is less significant. Indeed, it is clear from these two figures that on the grounds of independent evidence for identity, the highly-polar stationary phase DEGS makes an admirable complement to the nonpolar phases.

This point can be carried a stage further. R_F values in different paper chromatographic systems are not independent parameters, nor are R_F

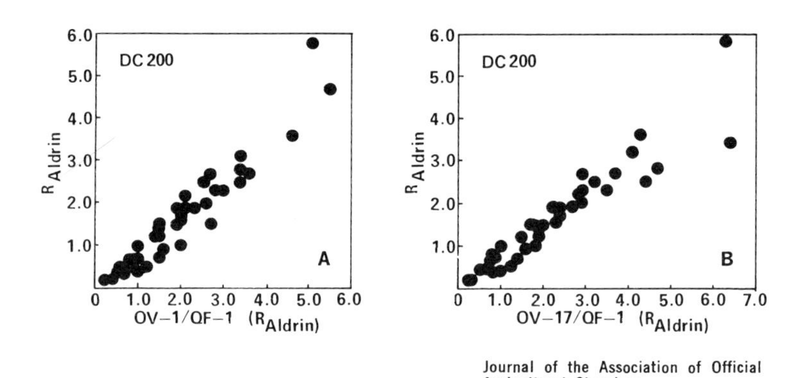

Journal of the Association of Official Agricultural Chemists

Figure 4. Gas–liquid chromatography; Ref. 11

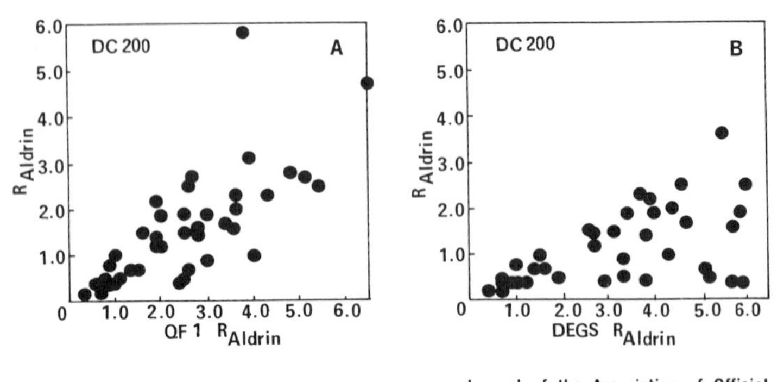

Figure 5. Gas–liquid chromatography; Ref. 11

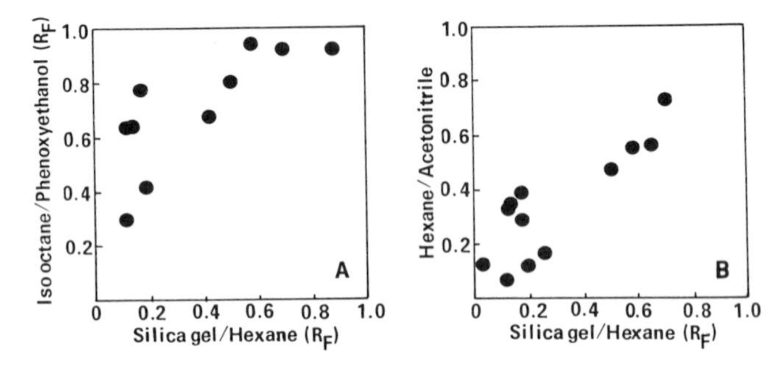

Figure 6. A) Paper chromatography vs. *thin-layer chromatography; B) Extraction* p-*value* vs. *thin-layer chromatography*

values in different TLC systems or extraction p-values in different pairs of solvents. In addition, since the basis of separation for all three of these techniques lies in partition phenomena, it would not be surprising to find some concordance between values for a compound by all these techniques. Figure 6 shows a plot of R_F values for organochlorine insecticides in a paper chromatography system against a thin-layer chromatography system and p-values against R_F in thin-layer chromatography. While an R_F from TLC gives some extra independent evidence over and above an R_F from paper chromatography, it is not great, and it is probably not worth the effort to use both methods; similarly for the p-value and the R_F on thin-layer chromatography.

By contrast, some techniques obviously depend on different physical effects to achieve separation. An R_F value obtained by paper chromatography would not be expected to correspond to a retention time on a GLC column since the former depends on partition and the latter, largely, on vapor pressure. Figure 7 shows plots of GLC *vs.* paper chromatography and also GLC *vs.* thin-layer chromatography. This figure shows that paper or thin-layer chromatography gives extra support to a tentative identification by GLC.

The point that has been made so far from a theoretical standpoint has practical meaning in the laboratory. Figure 8 shows GLC chromatograms for standard solutions of *p,p′*-DDE, *o,p′*-DDT, and *p,p′*-DDT, for a standard solution of polychlorinated biphenyls and of an extract of the liver of coho salmon, both concentrated and diluted (*12*). These were

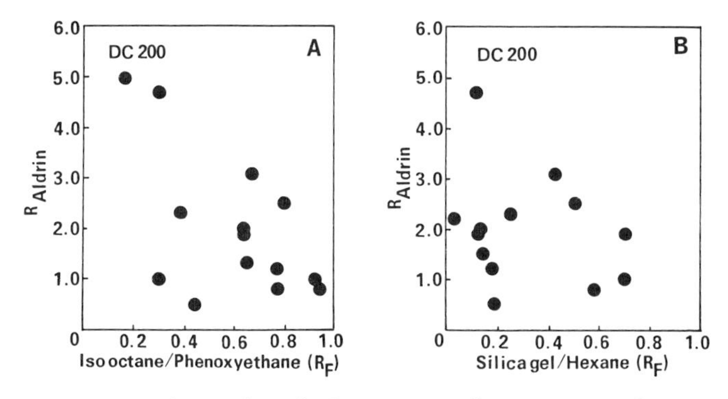

Figure 7. A) Gas–liquid chromatography vs. *paper chromatography; B) Gas–liquid chromatography* vs. *thin-layer chromatography*

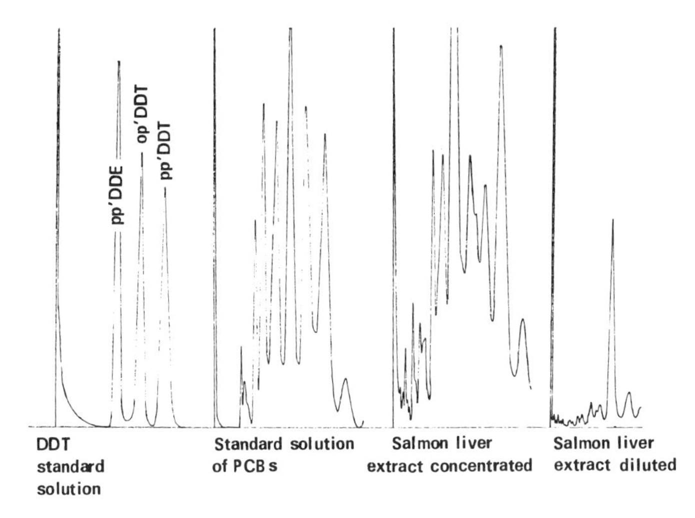

Figure 8. Gas–liquid chromatography of salmon liver extracts on SE 30

run on the stationary phase SE 30. However, the liver was shown by GLC/mass spectroscopy to contain not only DDE but also a constituent of polychlorinated biphenyls tentatively identified as pentachlorobiphenyl (*13*), and this has a retention time on SE 30 coincident with *p,p′*-DDE. Figure 9 shows the same extracts run on a polar stationary phase, phenyldiethanolamine succinate (PDES). The peak in the diluted liver extract contains both *p,p′*-DDE and pentachlorobiphenyl. This coincidence of retention times occurred on other polar stationary phases, such as the nitrile silicone XE 60. In other words, if the analysis of this extract had been carried out by GLC only, albeit on different stationary phases, the concentration of the DDE part of the mixed residue would have been incorrect.

Thus, techniques used to give further evidence for the identity of a pesticide residue should be intelligently chosen. If evidence from infrared or mass spectroscopy is not available, then adequate criteria of identification would be one or two GLC retention times, plus an R_F value from paper or thin-layer chromatography or an extraction *p*-value. Alternatively, one or two GLC retention times plus a GLC retention time of a derivative formed by chemical reaction would be a basis for confidence.

Much more emphasis must be placed upon this subject of the identification of trace constituents. The problem that the polychlorinated biphenyls present in the analysis of organochlorine pesticides has been illustrated. This is just one example of the need to identify accurately a very wide range of chemical compounds in extremely small concentrations.

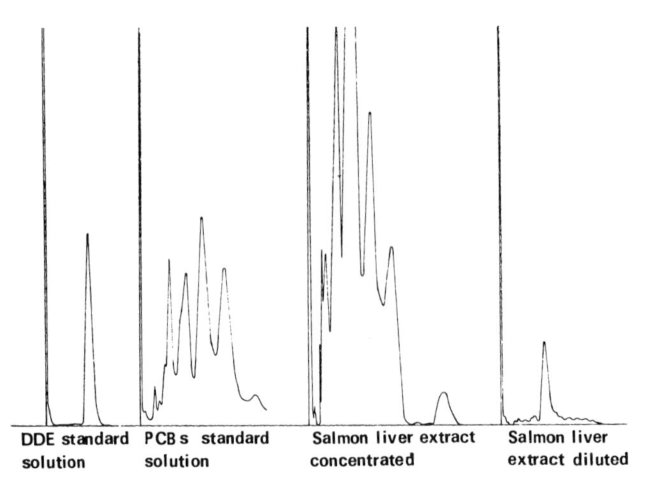

Figure 9. Gas–liquid chromatography of salmon liver extracts on PDES

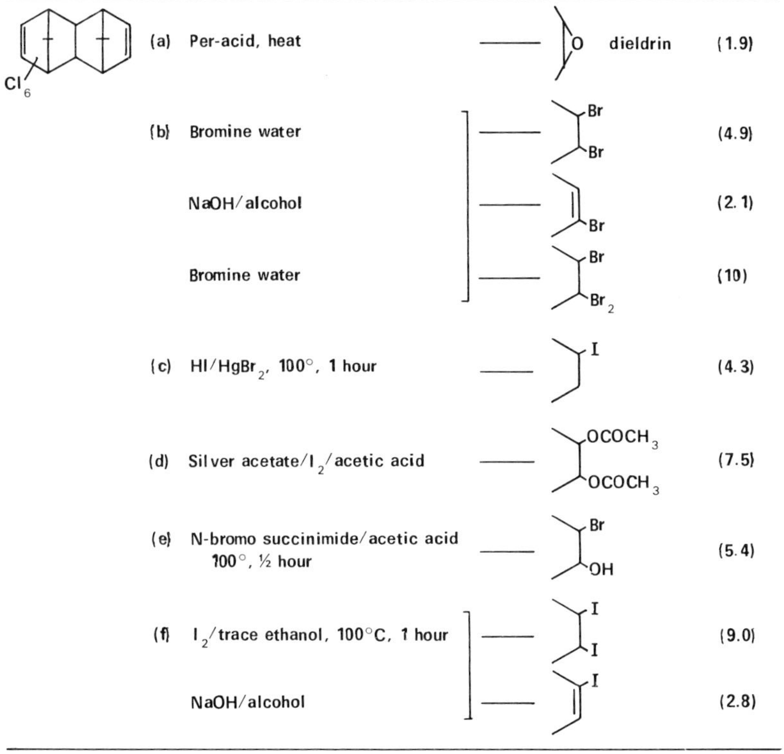

Cl_6

Reaction	Product	Retention time
(a) Per-acid, heat	O dieldrin	(1.9)
(b) Bromine water	Br, Br	(4.9)
NaOH/alcohol	Br	(2.1)
Bromine water	Br, Br_2	(10)
(c) HI/$HgBr_2$, 100°, 1 hour	I	(4.3)
(d) Silver acetate/I_2/acetic acid	$OCOCH_3$, $OCOCH_3$	(7.5)
(e) N-bromo succinimide/acetic acid 100°, ½ hour	Br, OH	(5.4)
(f) I_2/trace ethanol, 100°C, 1 hour	I, I	(9.0)
NaOH/alcohol	I	(2.8)

Bracketed numbers refer to GLC retention times on SE 30 relative to aldrin

Figure 10. Derivatives of aldrin

The emphasis now being placed upon the GLC/mass spectroscopy combination is already paying a rich dividend in some areas of analytical chemistry. Further research on this combination is bound to lead to a quickening interest from instrument manufacturers. The need is for an apparatus which is sufficiently sensitive and which, in terms of expense and expert labor, is within the range of laboratories specializing in residue analysis. When a reasonably inexpensive and easy-to-use equipment, capable of working at trace concentrations, becomes available, it will have the capacity to solve many analytical problems. However, it will be some years before this kind of instrument is on the market.

In addition to this approach, there is a need for fairly simple, routine tests that can be incorporated into the analytical procedures that every residue chemist can use. Confirmation procedures such as the formation of derivatives by chemical reaction were first applied some years ago, but it is only recently that Cochrane and Chau in Canada (*14, 15, 16, 17,*

18, 19, 20) and others (*21*) have begun to look at this approach systematically. The GLC of derivatives for organochlorine insecticide residues has been used in our laboratory since it was first appreciated how nonspecific the electron capture detector could be. Any derivative should retain as much as possible of the electron capture response; it should have a retention time different from that of the parent molecule, preferably a longer retention time in order to differentiate it clearly from the background. Figure 10 shows some work on aldrin with a view to finding suitable derivatives for confirmation. Our normal procedure for confirming any aldrin residue was to convert it to dieldrin with per-acid, a technique published by Noren (*22*) and by Osadchuk and Wanless (*23*). But Figure 10 indicates that there are a number of other derivatives which could be used for which the reaction is easy to carry out, gives a reasonable yield of product, and where the GLC peak is convenient for confirmation. Figure 11 shows some derivatives of dieldrin. Our usual confirmation was reacting the residue with acetic anhydride/hydrobromic acid, part of the old phenyl azide colorimetric method for dieldrin. At 100°C, the product is the bromoacetate derivative, but in the cold, as published by Hamence *et al.* (*24*), a mixture of bromoacetate and bromo-

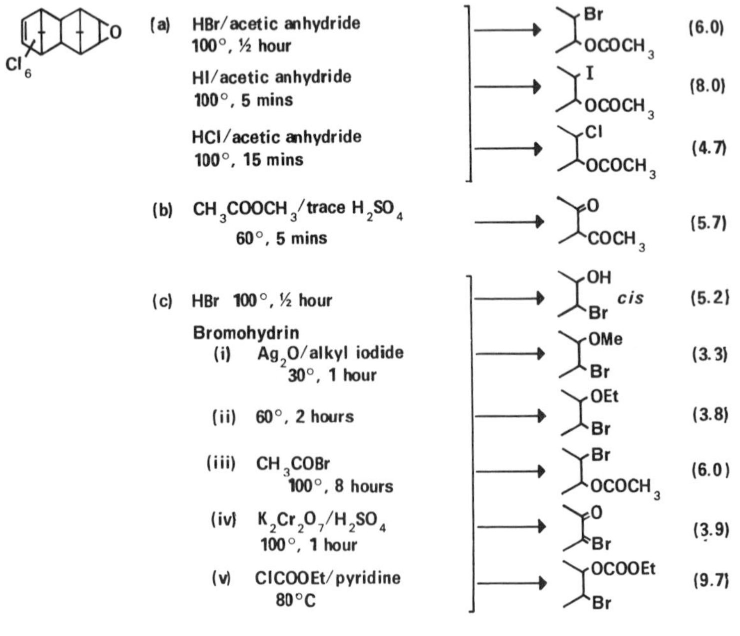

Cl_6 O

(a) HBr/acetic anhydride 100°, ½ hour → Br / $OCOCH_3$ (6.0)
HI/acetic anhydride 100°, 5 mins → I / $OCOCH_3$ (8.0)
HCl/acetic anhydride 100°, 15 mins → Cl / $OCOCH_3$ (4.7)

(b) CH_3COOCH_3/trace H_2SO_4 60°, 5 mins → O / $COCH_3$ (5.7)

(c) HBr 100°, ½ hour → OH / Br *cis* (5.2)
Bromohydrin
(i) Ag_2O/alkyl iodide 30°, 1 hour → OMe / Br (3.3)
(ii) 60°, 2 hours → OEt / Br (3.8)
(iii) CH_3COBr 100°, 8 hours → Br / $OCOCH_3$ (6.0)
(iv) $K_2Cr_2O_7/H_2SO_4$ 100°, 1 hour → O / Br (3.9)
(v) ClCOOEt/pyridine 80°C → OCOOEt / Br (9.7)

Bracketed numbers refer to GLC retention times on SE 30 relative to aldrin

Figure 11. Derivatives of dieldrin

hydrin is obtained. Again, a number of interesting derivatives were found which could be used for confirmatory purposes. By contrast, with endrin it seems difficult to make any other derivative than the pentacyclic ketone product, the so-called "Δ-keto," whatever reagents and conditions are employed.

One further way of improving the specificity of the analysis step is to remove more of the crop extractives that interfere with the analysis. The most commonly used are still partition between immiscible solvents and column chromatography. Although these remove most of the unwanted materials, techniques which have a different basis for separation have not yet been studied adequately. A separation based on molecular weight differences using gel permeation chromatography has been published by Ruzicka *et al.* (*25*) and by Horler (*26*). This technique warrants further study to find whether improved efficiency and resolution can be obtained. Robust permeable membranes are now marketed and these also have obvious applications to clean-up which should be studied. In fact, the whole area of interfering substances is still open to further research. Foodstuffs can be roughly classed as lipid, carbohydrate, and protein, plus other substances such as pigments, alkaloids, and minerals which are present in smaller concentration. Of these, the class which clearly represents the greatest contribution to GLC interference is the lipid. Included in the lipids are groups such as fatty acids, glycerides, phospholipids, resins, resin acids, plant sterols, terpenes, including carotenes and xanthophylls, and waxes which are mainly esters but which contain hydrocarbons, alcohols, ketones, and acids. As with the preparation of derivatives for confirmation of a residue, it appears that here is another area where the analyst might apply chemical reactions profitably. Lead acetate complexes or urea complexes with fatty acids could be used to retain selectively at least the saturated members of this group. Phosphatides are insoluble in acetone and some form complexes with magnesium or cadmium chloride. Hydrocarbons form picrates, and polyhydroxy compounds form borates. It may be that there is little which is novel in this approach and that while it is possible to remove interfering compounds as complexes on a chromatography column, the job can be done more easily another way. However, this approach of systematically clearing an extract of the major components of GLC interference seems never to have been attempted in pesticide residue analysis.

More attention should be paid to the needs of the residue analyst working at the bench. Multidetection analytical methods are an enormous gain, but we need to beware of constantly adding new pesticides or breakdown products to an already-existing multidetection scheme. To do this will eventually make confirmation of identity an impossible job. For example, the multidetection scheme based on electron-capture GLC for the organochlorine insecticides should be retained for this group and a separate scheme used for the organophosphorus compounds based on GLC with a phosphorus-sensitive detector. Separation in groups by column chromatography, already used for organochlorine insecticides and other pesticides by, for instance, Laws and Webley (*3*), McLeod *et al.* (*27*), and Sans (*28*) needs further study. Each group should have its own multidetection procedure, and confirmatory steps should be part of the procedure. Such separation schemes will be increasingly necessary as the emphasis in residue analysis moves towards environmental samples which may well contain contaminating substances from industry and commerce.

The confirmation of the presence of a residue often is carried out by people with little time or effort to spare. Apparatus and materials reduced to the small scale save time and do not necessarily sacrifice accuracy and precision. The accent must be on reliability and reproducibility, on speed and on simplicity.

Literature Cited

(1) Robinson, J., Richardson, A., Elgar, K. E., *ACS Meeting, New York, September 1966*, Abstracts of Papers A 075.
(2) Schechter, M. S., *Pestic. Monit. J.* (1968) **2,** 1.
(3) Laws, E. Q., Webley, D. J., *Analyst* (1961) **86,** 249.
(4) Blinn, R. C., ADVAN. CHEM. SER. (1971) **104,** 81.
(5) Biros, F. J., Walker, A. C., *J. Agr. Food Chem.* (1970) **18,** 425.
(6) Biros, F. J., ADVAN. CHEM. SER. (1971) **104,** 132.
(7) Mitchell, L. C., *J. Assoc. Offic. Agr. Chemists* (1958) **41,** 781.
(8) McKinley, W. P., Mahon, J. H., *J. Assoc. Offic. Agr. Chemists* (1959) **42,** 725.
(9) Abbott, D. C., Egan, H., Thomson, J., *J. Chromatog.* (1964) **16,** 481.
(10) Bowman, M. C., Beroza, M., *J. Assoc. Offic. Agr. Chemists* (1965) **48,** 943.
(11) Thompson, J. F., Walker, A. C., Moseman, R. F., *J. Assoc. Offic. Anal. Chemists* (1969) **52,** 1263.
(12) Richardson, A., Shell Research Ltd., unpublished work, 1970.
(13) Richardson, A., *ACS/CIC Meeting, Toronto, May 1970.*
(14) Cochrane, W. P., Chau, A. S. Y., *J. Assoc. Offic. Anal. Chemists* (1968) **51,** 1267.
(15) Chau, A. S. Y., Cochrane, W. P., *J. Assoc. Offic. Anal. Chemists* (1969) **52,** 1092.
(16) Cochrane, W. P., *J. Assoc. Offic. Anal. Chemists* (1969) **52,** 1100.
(17) Chau, A. S. Y., Cochrane, W. P., *J. Assoc. Offic. Anal. Chemists* (1969) **52,** 1220.
(18) Chau, A. S. Y., *J. Assoc. Offic. Anal. Chemists* (1969) **52,** 1240.
(19) Chau, A. S. Y., Cochrane, W. P., *Bull. Environ. Contam. Toxicol.* (1970) **5,** 133.
(20) Cochrane, W. P., Chau, A. S. Y., ADVAN. CHEM. SER. (1971) **104,** 11.
(21) Wiencke, W. W., Burke, J. A., *J. Assoc. Offic. Anal. Chemists* (1969) **52,** 1277.

(22) Noren, K., *Analyst* (1968) **93,** 39.
(23) Osadchuk, M., Wanless, E. B., *J. Assoc. Offic. Anal. Chemists* (1968) **51,** 1264.
(24) Hamence, J. H., Hall, P. S., Caverly, D. J., *Analyst* (1965) **90,** 649.
(25) Ruzicka, J. H., Thomson, J., Wheals, B. B., Wood, N. F., *J. Chromatog.* (1968) **34,** 14.
(26) Horler, D. F., *J. Sci. Food Agr.* (1968) **19,** 229.
(27) McLeod, H. A., Mendoza, C., Wales, P., McKinley, W. P., *J. Assoc. Offic. Anal. Chemists* (1967) **50,** 1216.
(28) Sans, W. W., *J. Agr. Food Chem.* (1967) **15,** 192.

RECEIVED July 16, 1970.

11

Analysis of Pesticide Residues: Immunological Techniques

C. D. ERCEGOVICH

Department of Entomology and Pesticide Research Laboratory, Pennsylvania State University, University Park, Pa. 16802

Past investigations which are known to have been conducted on immunological methods for analyzing pesticides and their degradation products are limited to those of only three separate research groups. Two of these groups were successful in developing antisera specific for DDA, a metabolite of DDT, and malathion. Parathion, aminotriazole, and several degradation products were readily detected by immunological methods by the third group of investigators. None had shown the practicality of such methods for the analysis of actual or fortified residue samples; however, their results clearly demonstrate a potential usefulness of immunological methods for detecting pesticides. The development of immunological methods for pesticide analysis as supplemental methods for rapid screening and confirmatory test is encouraged. The advantages and disadvantages of such methods are discussed.

In keeping with the theme of this volume, identification of pesticides at the residue (submicrogram) level, the questions of what applications of immunological techniques have been used in the past, and what potential usefulness do these techniques have for these purposes in the future are considered. The topic considers both immunological and serological techniques. Immunology, in a restricted sense, deals with the procedures used and the mechanisms involved whereby a host establishes resistance to a disease (immune state) after a specific exposure to a foreign infectious agent (antigen). In a broader sense, immunology is concerned with hypersensitive biological phenomena of altered tissue reactivity such as allergies, acquired tolerances to and rejection of foreign tissue, and autoimmune diseases. Serology is a branch of biological science concerned with the diagnostic and experimental procedures by which antigen-antibody reactions can be studied. It is so named because serology involves the use of serum. Since many immune reactions are studied by serological techniques, it is permissible to refer to the topic merely as immunological techniques.

The average pesticide residue chemist or toxicologist may have no or only a superficial knowledge of immunology and serology. Perhaps his familiarity with these sciences is limited to a sophomoric knowledge of their relation to immunization in disease prevention, diagnosis for allergies, blood typing, and disease resistance. Certainly these are among the primary concerns of immunology and serology, and more recently the importance of immunization as it applies to the rejection of organ transplants has given special importance to these fields. Because these sciences are highly specialized, it may be worthwhile to review briefly some of their underlying principles to assist better the pesticide researcher to gain an understanding and appreciation of how these techniques may apply to his interests. Since only a limited discussion of basic principles and methodology can be presented here, the reader is referred to excellent monographs by Burrows *et al.* (*1*), Day (*2*), Gary (*3*), and Weiser *et al.* (*4*) for the general principles of immunology, immunochemistry, and immune reactions, Landsteiner (*5*), Nezlin (*6*), and Pressman and Grossberg (*7*) for a better understanding about the biochemistry and specificity of antibodies, and Campbell *et al.* (*8*), Kabat and Mayer (*9*), Kwapinski (*10*), Nowotny (*11*), and Williams and Chase (*12*) for specific and detailed information about methodology.

General Principles

Essentially, when a foreign protein, antigen, is injected into a suitable animal, the organism responds by producing a specific protein called an antibody. The immunological response of an animal to the initial injection of an antigenic substance is not immediate, but after a suitable incubation period the properties of the blood serum in respect to the antigen are altered. This effect is demonstrated by the production of an immune serum, antiserum, which differs from normal serum in that it reacts either *in vivo* or *in vitro* with the homologous antigen. This property of antiserum is the result of the presence of antibodies, substances which are formed by the organism in response to the injected antigen. The antibody is found in the largest amount and most convenient form in the blood serum and is intimately associated with the serum protein. Antiserum may contain from 1 to 5 or even more, mg per ml of antibody protein. The antibody activity generally is localized in the globulin fraction of the serum protein. The immune globulin, however, is indistinguishable from normal serum globulin by chemical analysis. Antibody activity of antiserum may be destroyed by denaturation with heat, alcohol, or urea and by degradation with strong acid or alkali. Immune globulin reacts with substances such as ^{131}I or fluorescent dyes to label the antibody for various experimental purposes without loss of activity.

Although the stimulation of antibody production is usually observed with proteins, certain polysaccharides, such as the capsular components of pneumococci and other microorganisms, are also antigenic. The antigenic property of substances is variable in that some antigens give marked immune response while others stimulate only a low grade immune response as demonstrated by the small amount of antibody production. Substances such as ovalbumin, serum globulin, diphtheria toxin, and tetanus toxoids are examples of good antigens while hemoglobin, nucleoproteins, and histones are poor antigens. Good antigens are generally naturally occurring substances of relatively large molecular size, at least partially digestible by enzymes, and are foreign or contain structures which are foreign to the antibody-producing animal.

The naturally occurring proteins which are good antigens usually contain a full complement of amino acids. A number of derived proteins, notably gelatin, are not antigenic, or only feebly antigenic, even though they may be of high molecular weight. The poor antigenic property of gelatin is thought to be caused by its deficiency in aromatic amino acids, though there is little definitive evidence for this belief. Large molecular size invariably accompanies the ability to stimulate antibody formation, though all large molecules are not antigenic. There is some correlation between relative antigenicity and molecular weight since naturally occurring substances having molecular weights of less than about 40,000, *e.g.* lysozyme, protamines, histones, and insulin, are poor antigens. Consistent with this information the antigenicity of a protein disappears rapidly upon enzymatic hydrolysis, and proteoses and polypeptides are not antigenic. It is possible that high molecular weight synthetic polymers are antigenic, but with the exception of polyvinyl pyrrolidone none are found to be antigenic.

A prerequisite for antigenicity is that the antigen be foreign and distinguishable as "not-self" by the antibody-producing organisms. The foreign quality of antigens is of varying degree and is reflected in the specificity of the antigen-antibody reaction. Widely different antigens appear to be quite unrelated immunologically while closely related antigens show cross-reaction of varying degree with appropriate heterologous antibody and behave as poor antigens when they are closely related to constituent antigens of the immunized animal. Antigenic substances, therefore, from one species of animal stimulate antibody formation in another animal species, and antigens of plant origin are antigenic in animals.

This specificity is readily illustrated by the following example. Serum proteins, which are among the most commonly used antigens, from sheep, horse, chicken, man, etc., but not homologous rabbit serum, are antigenic in rabbit. Similarly the serum protein from rabbit, sheep, horse and man, but not the homologous chicken serum protein, are antigenic in chicken. Specificity is also expressed in the reaction of antigen with antibody. For instance, antiserum to horse serum protein prepared in the rabbit will react *in vitro* only with the homologous horse serum protein antigen and will not react with serum protein from chicken, cattle, man, rabbit, etc. Specificity of this nature is referred to as species specificity. On the other hand, specificity is less sharp between antigens from closely related sources. The antiserum to chicken serum protein will also react with pigeon serum protein; antiserum to sheep globulin will react with beef serum globulin, etc., but the heterologous reaction is weaker than that of the homologous antigen.

Experimental evidence established that the specificity of antigens is determined by their chemical composition. Experiments with a variety of antigenic proteins showed that immunologically identical proteins are, as far as can be determined, identical in composition. Antigenic proteins differing from one another in composition are also immunologically distinct while antigens showing some degree of cross-reaction are closely related in chemical structure. Through the study of altered specificity and artificial antigens evidence has been gained that immunological specificity is a property of molecular configuration. Antigenic protein may be heated, partially denatured, or treated with formaldehyde in such a way that part of the original specificity is lost, but species specificity remains, although somewhat broadened. Treatment of protein with iodine, nitric acid, or nitrous acid alters the specificity of the antigen so profoundly that species specificity is destroyed.

Landsteiner and colleagues (5) have shown that species specificity of antigens may be altered in ways other than by attacking the aromatic moieties of protein as with iodine and acids. The addition of small radicals, acetyl, ethyl, or methyl, to large protein molecules by acetylation with acetic anhydride, esterification with ethyl alcohol, or methylation with diazomethane, respectively, results in pronounced changes in specificity of the original protein molecule. The immunological properties of egg albumin have been altered through phosphorylation.

Up to this point we have been concerned primarily with so-called complete antigens, substances which both stimulate antibody production and react with the antibody so formed. Other substances, however, have a more limited antigenicity—*e.g.*, while they react specifically with antibody, they are unable to stimulate antibody formation. These substances are partial antigens or haptens. The use of immunological and serological techniques to detect and analyze pesticides depends on the understanding of haptens. Haptens are subdivided into two groups by some workers. One of these is made up of haptens which react with antibody *in vitro* to give the usual serological reactions; the other group includes those substances which react with antibody, but without overt evidence of the reaction, and the antigen-antibody reaction is demonstrable only indirectly as an interference or inhibition phenomenon. The latter is a consequence of the structure of the hapten and is not related to its antigenic specificity. The definition of a hapten has varied since the word was first used by Landsteiner in 1921 to refer to simple organic residues that react specifically with antibodies. However, none describe what is really meant as well as the original Greek meaning, to touch, to grasp, and to fasten (*2*). By simple immunobiological definition haptens are described as partial antigens which in themselves are incapable of inducing antibody formation in animals, but which, when attached to ordinary immunogens such as proteins and polysaccharides, induce antibody formation against themselves. Other substances—such as collodion particles, sephadex, red blood cells, and charcoal—have been used as the carrier material for haptens. Haptens range in size from small organic compounds such as *p*-aminobenzoic acid (*5, 13*) to polypeptide chains (*14, 15*).

The influence of the addition of relatively simple organic radicals on the immunological specificity of antigenic proteins has been explained by the study of the immunological behavior of a variety of compounds prepared from protein and the diazonium derivatives of a number of methyl, chloro, bromo, and nitro substitution products of aniline, *o*-, *m*-, and *p*-aminobenzenesulfonic acid, *p*-aminophenylarsenic acid, and *o*-, *m*-, and *p*-aminocinnamic acid as well are the parent compound (*16, 5, 13*). From studies of this nature it became apparent that the immunological behavior of antigens can be modified specifically by altering a relatively small part of the large protein molecule and that the specificity of the antigen is determined by the chemical structure of this added part. The spatial arrangement in the determinative groups, as well as their nature, is reflected in immunological behavior. The position of ortho, meta, or para substitutions in aromatic radicals attached to proteins produces differences in specificity. The stereo isomers of tartaric acid and *p*-aminobenzoylphenylacetic acid yield immunologically distinct antigens when coupled with protein, and the position of amino acids in peptide-azoproteins is a factor in determining immunological specificity.

The immunological functions of the haptens now are clear in their role as determinants of immunological specificity. The relative importance of hapten-determined specificity is very great since a large portion of naturally occurring antigenic substances are conjugated antigens consisting of a protein and a hapten. This is illustrated for some pneumococcus whose polysaccharide capsular substance acts as a hapten and provides the antigenic specificity required to produce an immunity to the injected microorganisms. Immune response, often manifested as a hypersensitivity, to low molecular weight substances such as drugs, dyes, cutting oils, and similar substances responsible for occupational dermatitis, occur frequently. It is probable that these low-molecular-weight substances function as a hapten by combining with the host's protein, thus altering their specificity so that they become foreign, antigenic substances which provoke immune responses. The antibody formed reacts specifically with the haptenic substance producing allergic reactions. Sensitization to penicillin, which is relatively often observed, is an excellent example of this phenomenon.

Past Attempted Uses of Immunological Techniques for Pesticide Analysis

Immunological techniques have been used extensively in the field of medicine to study the causes and treatment of diseases. The protein and enzyme chemist has also used immunological methods to good advantage to isolate, purify, and study proteins. In agriculture such methods have been used by the virologist who uses antigen-antibody reactions to diagnose plant diseases. The veterinarian uses these techniques to diagnose animal diseases and to identify pathogenic bacteria and viruses. Entomologists have used immunological methods to study the venom of insects and to develop ways of immunizing against such substances. They have also used immunospecificity as a criterion to study the differences in various insect enzymes—*e.g.*, acetyl cholinesterase and DDT-dehydrochlorinase—to understand the mode of action of insecticides and the mechanism of insect resistance to insecticides.

Attempts to utilize the specificity of antigen–antibody reaction for the analysis of pesticides appear to have been limited to work of Centeno *et al.* (*17*) and Haas and Guardia (*18*). The former group postulated, because of observations of allergic contact sensitivity to malathion and scattered reports about unusual reactions to DDT, that common pesticides might act as antigenic and allergic determinants, provided they become coupled to an appropriate protein carrier. To test this postulate, they studied the immunogenicity of protein conjugates of closely related derivatives of DDT and malathion, the metabolite DDA, [2,2-bis(*p*-chlorophenyl) acetic acid], and [O,O-dimethyl S-(1,2-biscarboxyethyl) phosphorodithioate]. Their results were highly significant because they were able to produce antibodies in rabbits to the above haptens when conjugated to bovine serum albumin (BSA), DDA-BSA, and malathion-BSA, respectively.

A method for ascertaining the antibody titer of the respective antisera was developed using a bis(diazotized–benzidine) hemagglutination system. Though significant titers to these conjugates were detected in all of the animals, the conjugated malathion antigen elicited antibody response in much higher titer and after fewer injections than did the DDA antigen. Gel diffusion tests established the specificity of reaction between antisera to DDA-BSA and malathion-BSA with their respective homologous BSA antigen. The antisera to malathion-BSA gave no detectable precipitin reaction with bovine serum albumin, indicating that the carrier protein underwent marked configurational changes during conjugation or that some of its antigenic determinant groups were masked. Centeno *et al.* (*17*) also stress that the DDA-BSA and malathion-BSA conjugates did not react with antisera to bovine serum albumin, indicating that they were antigenically different from the original carrier protein. Their study demonstrated that antibodies to DDT and malathion metabolites are readily produced, and they suggest that these antibodies serve for the development of sensitive histological procedures along with radioimmunochemical and fluorescein labelling techniques, to determine the localization of these chemicals in animal tissues.

The report by Haas and Guardia (*18*) pertains to their efforts to apply immunological methods for the assay of pesticide residues, and expresses their aim to test the suitability of methods for field analytical purposes. Haas and Guardia also used DDT and malathion to represent two of the most important classes of insecticides, chlorinated hydrocarbons and organophosphorus compounds. They first attempted to prepare insecticide–protein antigens in which enzymes were used as the protein carrier. Antiserum of rabbits injected with DDA-carbonic anhydrase or malathion-chymotrypsin failed to show the presence of the respective antibodies.

They experimented with four other proteins as carriers: rabbit serum albumin, bovine serum albumin, bovine fibrinogen fraction I, and bovine β-globulin fraction III. The structurally related derivatives of DDT and malathion, DDA, and *O,O*-dimethyl S-carboxy-carboxyethyl phosphorodithioate (malathion half ester), respectively, were used as the specific haptens attached to these carrier proteins. These compounds contain free carboxyl groups, which when they reacted with thionylchloride, provide a means of coupling of the hapten to the amino groups of the protein carrier.

Unlike Centeno *et al.* (*17*) Haas and Guardia could not show the presence of antibodies in the antiserum of rabbits immunized with the hapten conjugates of bovine serum albumin. The results with β-globulin conjugates were inconsistent, whereas the DDA and malathion half-ester conjugates with fibrinogen gave the best antigenic responses. Through the serological test methods of precipitation, tanned cell hemagglutination, and tanned cell hemagglutination inhibition they demonstrated the presence of antibodies to DDA-fibrinogen and malathion-fibrinogen in the antiserum of the respectively immunized rabbits.

Using the tanned cell hemagglutination inhibition test, they determined that the sensitivity of this method was between 0.1 and 1.0 μgram for DDA and malathion. DDT could not be detected in this manner. Neither DDT, DDA, nor malathion could be detected by direct reaction in the precipitin or tanned cell hemagglutination tests. The antibodies produced against the various conjugates were specific for the particular protein carrier since the immune serum from rabbits inoculated with DDA-fibrinogen did not agglutinate the DDA-rabbit serum albumin conjugate and vice versa. Antihapten antibodies appeared to be transitory. Maximum titer was obtained approximately 6 weeks after the initial inoculation and no reaction was detectable 3 weeks later. The haptenic antisera were unstable after a few days of storage at 4° or −10°C. The antibody activity appeared to be maintained if the antisera were first frozen with dry ice and stored at −30°C.

Prior to learning about the aforementioned groups' efforts investigations were begun in the author's laboratory to test the usefulness of immunological methods for detecting and analyzing pesticides and pesticidal degradation products in biological specimens. Only a summary of these studies can be presented here since our findings are considered preliminary and specific details will be published subsequently elsewhere.

Unlike the other workers who selected compounds possessing carbonyl groups for coupling purposes to the protein carriers, our investigations centered on pesticide compounds which possess either an amine group or a nitro group, which can be reduced readily to an amine. These included the herbicide aminotriazole, the insecticide parathion, and pesticide degradation products nitrophenol, 4-chloronitrotoluene, and aniline. Hapten–protein conjugates of all of these materials were readily prepared by modifying slightly the method described by Williams and Chase (*12*). An unsuccessful attempt to use egg albumin as the carrier protein led to the use of bovine plasma protein. These carrier proteins were unsatisfactory because they lacked specificity and low titer formation in the antiserum. The high degree of heterogenous antibody production created problems of cross reaction which could not be corrected adequately through antiserum-antigen adsorption techniques. Following the undesirable results with these two protein carriers, bovine fibrinogen fraction I was tested for this purpose at the suggestion of Haas (personal communication) (*19*). The first results obtained from tests using fibrinogen as the protein carrier revealed a high degree of specificity toward the homologous azoprotein antigens. This specificity was somewhat greater, however, during the first hours of incubation and decreased as a function of time with a limiting value of approximately 24 hours. Very little, if any, cross-reactions were observed in the early hours of incubation, but heterogenous precipitates began to form after about eight hours of incubation of the precipitin tests. In a second immunization trial with the various hapten fibrinogen azoproteins, somewhat higher titers of antisera were produced but with a greater loss of specificity as evidenced from the decreased length of time involved for cross-reactions to develop.

In a third attempt to use fibrinogen as the carrier protein a different immunization schedule was followed. Whereas in the earlier experiments the animals, New Zealand male and female rabbits, were bled for antiserum recovery four days after the last of seven injections, intervals of two days after the last injection of the hapten-protein conjugates were tried. Re-immunization of the rabbits was repeated 30 days afterwards instead of the former six-day interval. Antisera of higher titer were produced in all cases, however, the shorter period of immunization did not prevent the production of relatively high titers of heterogenous antibodies as evidenced by the high degree of cross reactions. In addition to the lack of specificity and cross-reactions with the fibrinogen conjugates this protein was less desirable because of the difficulty experienced in the preparation of work solutions of its hapten-azoprotein conjugates. To dissolve such azoproteins a solution of 8*M* urea and sodium hydroxide was used. Urea, a mild denaturing agent, disrupts the secondary structure of proteins by causing the protein chain to become stretched, thus exposing an extremely large number of possible antigenic sites. Once injected into an animal such chains apparently are easily degraded into smaller fragments which are also capable of inducing antibody formation. This phenomenon may explain the lack of specificity and high production of heterogenous antibodies in our antisera as demonstrated by precipitin tests. It is virtually impossible to remove these interferring heterogenous antibodies from the antiserum by adsorption techniques with antigen since the responsible antigens were formed *in vivo* owing to degradation of the originally injected azoprotein. Adsorption with fibrinogen alone removed only a small fraction of the overall interferring heterogenous antibodies.

Bovine serum albumin proved to be a much better protein carrier of the haptens used in our investigation. This protein was also used by Centeno *et al.* (*17*) whose production of antisera to the DDA–BSA and malathion–BSA conjugates is discussed in preceding paragraphs. Bovine serum albumin is relatively smaller in molecular size than fibrinogen. The azoprotein conjugates produced with it can be solubilized by less erratic procedures which cause less structural alterations of the carrier protein, therefore producing fewer heterogenous antibodies. The extraneous antibodies which do occur when bovine serum albumin is used as the hapten carrier is removed readily from the antiserum by adsorption with the unconjugated protein.

A digest of the limited amount of information available about the subject causes this reviewer to be somewhat optimistic about the potential usefulness of immunological methods for the analysis of pesticides. Experiences gained from investigations conducted in our laboratory reveal that methodology exists which can be used to implement the development of immunological procedures for pesticide analysis. The three groups who are known to have tested immunological methods for detecting pesticides were all successful in developing antisera for some very common pesticides and their degradation products. Two of the groups demonstrated that it was possible to detect submicrogram quantities of these chemicals. The same workers have shown that while it may not be possible to detect these pesticides by direct serological tests—*e.g.*, precipitin and hemagglutination reactions—these chemicals can be detected in trace amounts by indirect serological tests which are relatively simple to conduct. Two of the most useful indirect methods which can be used to detect the unconjugated form of the pesticides are the hapten inhibition of precipitation and passive hemagglutination inhibition test. The principle of hapten inhibition involves the reaction of an antibody with free hapten and observing a decrease in the precipitation of the antibody with the addition of the appropriate antigen—*i.e.*, the hapten conjugated precipitating protein. The amount of precipitation is directly proportional to the amount of hapten present to compete for active sites on the antibody. The hapten inhibition of precipitation method is more specific than the passive hemagglutination inhibition method but is less adaptable for reliable quantitative data.

Prognosis for the Future Use of Immunological Methods for Pesticide Analysis

The pesticide chemist will question what the practical applications of immunological techniques for residue analysis of pesticides are. He will be concerned about the specificity, quantitative aspects, and advantages these techniques have over the existing colorimetric, chromatographic, spectrophotometric, and bioassay methods which work for him. Because of the stringent requirements to amass all of the necessary toxicological and residue data needed to obtain registration for the sale of a pesticide, one of the first objectives of a would-be producer is to develop reliable methods to analyze his product. For this reason there exist adequate methods for analyzing food products, components of the environment, and pathological specimens for specific pesticides and their degradation products. These methods, however good they are for analyzing specific compounds, are not adequate to analyze efficiently and meaningfully the general food supply, and all of the other samples of concern, for the comprehensive qualitative and quantitative information which is desired about the pesticides that may be present.

Pesticide laboratories of food industries and regulatory agencies are continually faced with the problem of analyzing samples whose history of exposure to pesticides is unknown. More than one pesticide may be present in any of these samples and the residue of each may have to be determined. To help solve this problem of analyzing diverse sample types for exposure to different types of pesticides, effort has been made by the FDA scientists, among others, to develop methods for the multiple analysis of pesticides.

The subject of methods for the multiple analysis of pesticides has been adequately reviewed recently by Burke (*20*). According to Burke, however, a definite procedure for confirming the identity of a given pesticide residue has not been established. Thin-layer chromatography, gas chromatography on columns that give different retention patterns, and the p-values of Bowman and Beroza (*21, 22*) are probably the most universally applicable confirmatory techniques. Derivatization of the residue in question followed by gas chromatography of the derivative seems an excellent approach, and several procedures are described in the recent literature (*23, 24*). Multiresidue methods have a serious limitation since many pesticide chemicals are not determined and not all sample types can be handled in a routine fashion. Also, as previously mentioned, there is no complete scheme for confirming the identity of many pesticides. The selection of methods of analysis, therefore, depends extensively upon the pesticides and nature of the sample to be analyzed, the equipment and personnel required, and the simplicity, speed, cost, accuracy, and reproducibility of the method. Time frequently plays a predominant role in the selection of methods to be used by the food processor, who must comply to pesticide tolerances, and regulatory agencies, who must enforce them. Apparently, a combination of screening and specific methods is needed by such organizations since neither types of methods alone can satisfy their analytical requirements.

Some techniques may offer selective screening as well as specificity —*e.g.*, microcoulometric methods described by Coulson *et al.* (*24*). This technique consists of a combination of gas chromatography, combustion, and continuous coulometric titration for chlorine or sulfur. The development of the flame photometric detector offers a similar potential for the selective screening and specificity of pesticides which contain phosphorus or sulfur (*25*). Even so, one or more tests in addition to the initial analysis may be required to identify properly the pesticides present in a sample.

A critical evaluation of the limited amount of information available about the detection of pesticides by immunological methods enables one, without too much optimism, to reply in the affirmative to the residue chemist's question regarding specificity. Both Centeno *et al.* and Haas and Guardia were successful in producing antisera which were specific for derivatives of DDT and malathion, DDA and malathion esters, respectively. Antisera which were specific for parathion, aminotriazole, *p*-nitrophenol (a degradation product of parathion), and 4-chloro-2-nitrotoluene have been produced in the author's laboratory. The productionof these antisera was accomplished by conventional immunological techniques. Confirmation of these antibodies was readily demonstrated by well known serological methods.

The relation between antigen and antibody is highly specific, as indicated. In a sense the serological methods used to demonstrate antigen-antibody reactions are analytical methods, sensitive, semi-quantitative, and highly specific. There is a limit, however, to the precision of serological specificity. When different homologous antigens are used to immunize animals, the antibodies for one antigen react also with other antigens, though less strongly. These are called cross-reactions. Saying that an antibody is specific for a particular antigen is a relative term because it is based on the degree to which an antibody reacts with various antigens, not requiring it to react solely with only one of them. It is also relative because the antibody may not have been tested against all possible antigens; thus there could be others with which it might react.

Antiserum to an azoprotein or to other types of conjugated protein is produced primarily to obtain antibodies against known structures. For some unknown reason haptenic groups are not very immunogenic, regardless of the protein carrier, and antiserum usually contains a relatively low concentration of antibody against the conjugated hapten. In working with conjugated proteins it should be remembered that antibodies produced to it may exhibit three types of specificity: one directed toward the hapten, one toward the protein carrier, and one toward the hapten-protein conjugate. As a rule an antiserum against a given conjugated protein contains a diversity of antibodies. Because of this mixture of antibody molecules, a serological test for hapten antibody must be made with the hapten attached to the protein molecule having no cross-reaction with the one used for immunization. The interferring antibodies, which give rise to cross-reactions, however, are removed from antiserum by a process known as adsorption. This process involves incubating the antiserum with respective antigens which are required to react with the extraneous antibodies. The resulting antigen–antibody complexes can then be removed by centrifugation, leaving a specificity-enriched antiserum.

The exploratory investigations which have been conducted using immunological and serological techniques to analyze pesticides have been restricted to atypical sample conditions. Through the use of tanned cell hemagglutination inhibition tests, Haas and Guardia could detect quantities of 0.1 and 1.0 μgram of DDA and malathion. However, by this same technique they were unable to detect DDT with the antiserum which was produced in response to immunization with DDA-fibrinogen. These findings are significant because they demonstrate specificity for one of the major degradation products of DDT and that such procedures can detect microgram quantities of two important pesticidally related chemicals. It is of interest to attempt to conjugate DDT itself to a carrier protein by making an arsenate derivative of it, or a compound more closely related structurally to DDT than DDA. Such a compound would be Kelthane, [1,1-bis(*p*-chlorophenyl)-2,2,2-trichloroethanol].

Preliminary trials using a similar type of passive hemagglutination techniques enabled Babish (personal communication) (*26*) to detect 12 ngrams of parathion and 150 ngrams of aminotriazole. Neither group of workers, however, attempted to detect any of these compounds in crude or fortified extracts of soil, plant, or animal tissue. Therefore, the question about the suitability and effectiveness of detecting residues of these pesticides in actual samples remains to be answered. Nonetheless, these findings suggest that a combination of immunological and serological techniques have the potential of quantitation that the residue chemist requires and should be examined more seriously for pesticide residue analysis.

There are a number of obvious advantages to the use of immunological techniques for pesticide analysis, but the converse is also true. No inference is made that such techniques could ever replace the more sophisticated methods of analysis presently being used. It is only suggested that they have potential to supplement rather than supplant existing methods of analysis. Their greatest usefulness is for the rapid screening of a large number of samples for the presence of specific types of pesticides or threshold quantities. These methods are also ideal for confirmatory tests. Through the use of proper methods a very high degree of specificity for various compounds could ultimately be developed. Such confirmatory tests would be independent of reliance upon R_f values, retention times, and other highly empirical indexes presently used in conjunction with gas, liquid, and thin layer chromatography. Perhaps one of the greatest potentials for immunological methods for pesticide residue analysis is their use in developing countries. Because no highly specialized or expensive equipment is required, except for a suitable centrifuge. Personnel requirements would not be critical, except for a senior member who has the proper training and understanding of immunological methods. The operating technicians would not have to be highly skilled in electronics and other disciplines which are required for the proper operation of gas chromatographic and spectrometric equipment.

The overall advantages of immunological methods for pesticide analysis are that they can be performed relatively fast, economically, and simply with a high degree of accuracy and reproducibility. Since no high cost elaborate equipment is required, there would be no great initial investment, subsequent replacement, and servicing costs. Technician requirements in training and specialization would be minimal as well as the number of personnel needed. In comparison with the 20 to 40 analyses which a technician can perform on one gas chromatograph during an eight-hour work day his productivity with serological methods could be increased many-fold.

Even though the problems of producing highly specific antisera can be overcome, these methods may still have certain disadvantages. The foremost disadvantage concerns the source and availability of the respective antibodies which would be needed for analysis. This problem is not insurmountable for it is not expected that each laboratory wishing to use such techniques would produce its own antisera, except in those few cases in which the volume and type of operation would justify doing so. If and when immunological techniques are proved reliable and worthy for the routine analysis of pesticide residues, one would expect that specific antisera would eventually become available from commercial sources as is now true for a wide variety of antisera. One supplier of antisera presently lists a number of antisera of the nature discussed in this chapter for the following haptens: 3-indoleacetic acid, gibberellic acid, dinitrophenol, azobenzenearsonate, thymidine, and uridine. Even in the low quantity in which these antisera must be produced, owing to their very limited use, their cost is competitive with present methods of analysis. Most of these antisera are produced in rabbits. Greater quantities at substantially lower prices could easily be produced to meet high volume demand by using larger animals such as goats, cattle, and horses. One might also expect that the commercial supplier of antisera would also supply identification kits which contain both the specific antigen and specific antibody as controls and for positive identification.

Antisera may be highly perishable substances in comparison with the chemicals and solvents normally used by the pesticide chemist. Haas and Guardia suggested that a problem existed in this respect. Their preliminary results showed that the antiserum to DDA–fibrinogen and malathion–fibrinogen could be preserved up to one month if frozen with dry ice and kept at −30°C. More work on the stability of such antibodies would have to be conducted, however, before generalizations about their shelf life can be made. Suppliers of a number of antisera claim shelf lives of up to one year if the product is maintained under the recommended storage conditions. The disadvantage in this respect may amount mostly to a matter of more critical planning in regards to needs and inventory.

Immunological methods would not have the quantitative precision of gas chromatographic, colorimetric, or spectrophotometric methods. However, in their proper manipulation they can provide semiquantitative data comparable with that obtained by thin layer chromatography. The qualitative information derived from immunological methods would not necessarily be absolute but certainly would be more reliable than such information based on retention time and R_f values. The qualitative aspects of immunological methods will depend upon specificity, which will be influenced by the quality of the antiserum and the nature of the antigen (the residue sample). In this regard the question about sample preparation and cleanup arises. Unfortunately, detailed studies in this connection have not been conducted. One reason for deferring such studies in preference to working with pure solutions has been to establish the feasibility of such methods and that on a theoretical basis one expects a lesser degree of interference with serological methods than with more conventional methods of analysis. The structures of most of the pesticidal compounds differ so greatly from the normal constituents of plant, soil, and animal tissues that it would be unlikely to expect many of these substances to fit the steric requirements of highly purified and specific antiserum. At most one would expect that cleanup procedures used for gas and thin layer chromatography would be sufficient for serological detection. However, if little or no cleanup at all would suffice, the productivity per technician would be greater.

A certain amount of specialized training will be required for other methods of analysis. Proper understanding of immunology and serology would be required of the senior investigator to interpret the results properly, but this requirement is no more of an imposition or deterrent for these techniques than it is for other methods of analysis.

Literature Cited

(1) Burrows, W., Moulder, J. W., Lewert, R. M., "Testbook of Microbiology," 18th ed., W. B. Saunders, Philadelphia, 1965.
(2) Day, E. D., "Foundations of Immunochemistry," Williams & Wilkins, Baltimore, 1966.
(3) Gary, D. F., "Immunology," 2nd ed., Cheshire, Melbourne, Australia, 1970.

(4) Weiser, R. S., Myrvik, Q. N., Pearsall, N. N., "Fundamentals of Immunology," Lea & Febiger, Philadelphia, 1970.
(5) Lansteiner, K., "The Specificity of Serological Reactions," Dover, New York, 1962.
(6) Nezlin, R. S., "Biochemistry of Antibodies," Plenum Press, New York, 1970.
(7) Pressman, D., Grossberg, A. L., "The Structural Basis of Antibody Specificity," New York, 1968.
(8) Campbell, H., Garvey, J. S., Cremer, N. E., Sussdorf, D. H., "Methods in Immunology," W. A. Benjamin, New York, 1964.
(9) Kabat, E. A., Mayer, M., "Experimental Immunochemistry," Charles Thomas, Springfield, 1966.
(10) Kwapinski, J. B., "Methods in Serological Research," Wiley & Sons, New York, 1965.
(11) Nowotny, A., "Basic Exercises in Immunochemistry," Springer-Verlag, New York, 1969.
(12) Williams, C. A., Chase, M. W., "Methods in Immunology and Immunochemistry," Academic Press, New York, 1967.
(13) Mattioli, C. A., Yazi, A., Pressman, B., *J. Immunol.* (1968) **101,** 939.
(14) Gill, J. T., Boty, P., *J. Biol. Chem.* (1961) **236,** 2677.
(15) Talano, R. C., Haber, E., Austen, K. F., *J. Immunol.* (1968) **101,** 333.
(16) Haurowitz, F. J., *Immunology* (1942) **43,** 311.
(17) Centeno, E. R., Johnson, W. J., Sehon, A. H., *Int. Arch. Allergy Appl. Immunol.* (1970) **37,** 1.
(18) Haas, G. J., Guardia, E. J., *Proc. Soc. Exp. Biol. Med.* (1968) **129,** 546.
(19) Haas, G. J., personal communication (1971).
(20) Burke, J. A., *Residue Rev.* (1971) **34,** 59.
(21) Beroza, M., Inscoe, M. N., Bowman, M. C., *Residue Rev.* (1969) **30,** 1.
(22) Bowman, M. C., Beroza, M., *J. Assoc. Offic. Agr. Chem.* (1965) **48,** 943.
(23) Chau, A. S. Y., Cochrane, W. P., *J. Assoc. Offic. Anal. Chem.* (1969) **52,** 1220.
(24) Coulson, D. M., Cavanagh, L. A., DeVries, J. E., Walther, B., *J. Agr. Food Chem.* (1960) **8,** 399.
(25) Brody, S. S., Chaney, J. E., *J. Gas Chromatogr.* (1966) **4,** 42.
(26) Babish, J. G., personal communication (1971).

RECEIVED July 16, 1971.

INDEX